7 D0586799

K F ... ARY

The Key to Earth History

Second Edition

The angular unconformity described by James Hutton at Siccar Point, Berwickshire. Here Hutton first realised the antiquity of the Earth [Drawing by H. Foxwell]

The Key to Earth History

An introduction to stratigraphy

Second Edition

PETER DOYLE, MATTHEW R. BENNETT
ALISTAIR N. BAXTER

School of Earth & Environmental Sciences, University of Greenwich, UK

JOHN WILEY & SONS, LTD

Chichester • New York • Weinheim • Brisbane • Singapore • Toronto

Other Wiley Editorial Offices

John Wiley & Sons, Inc., 605 Third Avenue,
New York, NY 10158-0012, USA

WILEY-VCH Verlag GmbH, Pappelallee 3,
D-69469 Weinheim, Germany

John Wiley & Sons Australia, Ltd, 33 Park Road, Milton,
Queensland 4064, Australia

John Wiley & Sons (Asia) Pte Ltd, 2 Clementi Loop #02-01,
Jin Xing Distripark, Singapore 129809

John Wiley & Sons (Canada) Ltd, 22 Worcester Road,
Rexdale, Ontario M9W 1L1, Canada

Library of Congress Cataloging-in-Publication Data
Doyle, Peter.
The key to earth history : an introduction to stratigraphy / Peter Doyle, Matthew R. Bennett, Alistair N. Baxter.—2nd ed.
p. cm.
Includes bibliographical references and index.
ISBN 0-471-49216-7 (alk. paper)—ISBN 0-471-49215-9 (alk. paper)
1. Geology, Stratigraphic.   I. Bennett, Matthew (Matthew R.) II. Baxter, Alistair N. III. Title.

QE651 .D64 2001
551.7—dc21   00-068516

British Library Cataloguing in Publication Data
A catalogue record for this book is available from the British Library

ISBN 0-471-49216-7 (Cloth)
ISBN 0-471-49215-9 (Paper)

Printed and bound by CPI Antony Rowe, Eastbourne

Contents

Preface

Six years on from the publication of our first edition, this book remains a personal view of a subject which has developed beyond recognition in recent years. In our original preface we expressed the view that stratigraphy has been traditionally considered, by students at least, to be boring; a simple recital of dates and rock units. Since then our subject has developed rapidly, and the application of new techniques, and rediscovery of old ones, has led to its renaissance. This edition takes into account these new developments and delivers them to an audience with the fundamental message that stratigraphy is the key to Earth history.

As with our first edition, this book was written by Peter Doyle and Matthew Bennett, with support and guidance from Alistair Baxter. We thank all those who assisted us in developing our original concept and delivering the first edition; but the burden of revision was shouldered by ourselves, born from 10 years of teaching the subject and a desire to update our original text in the light of recent developments. As such we alone are responsible for its conception, its broad scope, as well as its errors and omissions.

The new edition holds true to our original concept; that the book should provide an understanding of the pattern and process of Earth history. As before, we hope that it will generate interest and enthusiasm for a subject which is vital in understanding our planet. We dedicate the book to stratigraphers everywhere and to those individuals who found some value in our original edition; and, on a more personal level to *our* own next generation, James, Edward and Jeremy.

Peter Doyle
Matthew Bennett
Alistair Baxter
Chatham, 2000

Illustrations

Many of the illustrations are not original. Where they are reproduced exactly as they were first published, permission has been obtained from the copyright holder, and is indicated in the figure caption by the word *from* followed by the source. Every effort has been made to obtain permission from the relevant copyright holders, but where no reply was received we have assumed that there was no objection to our using the material. Where we have significantly altered an illustration, we have acknowledged this and signalled the change in the figure caption by using the phrase *modified from*.

We are grateful to the following for permission to reproduce copyright figures and photographs: The Natural History Museum (Figures 3.1, 3.6 and 4.10); Cambridge University Press (Figures 5.7 and 5.14); The Association of American Petroleum Geologists (Figure 5.18); Geological Society Publishing House (Figures 7.3 and 7.4); Merrill Publishing Company (Figure 10.12); the Palaeontological Association (Figure 10.14). Figures 3.2, 3.5, 3.7, 5.2C, 5.2D, 7.2 and 10.16 are reproduced by permission of the Director, British Geological Survey: NERC copyright reserved.

1

Introduction: What is Stratigraphy?

1.1 Stratigraphy: the Key to Earth History

Geology is the science of the Earth. Its aim is to understand the composition, structure and history of the Earth throughout the 4600 Ma of its existence. There are many different branches to geology: geochemistry, petrology and mineralogy are the branches concerned with the chemical and mineral composition of the Earth; sedimentology and geomorphology deal with the surface processes which shape the Earth's crust; geophysics looks at the structure of the Earth, both at the surface and at depth; palaeontology is concerned with the past life of the planet. These subjects have parallels with other physical sciences, but geology has the unique component of time: the development of the Earth in time and space. Within geology the study of time is the study of stratigraphy. Stratigraphy is the key to understanding the Earth's crust and its materials, structure and past life. It encompasses everything that has happened in the history of the planet. All geologists, whatever their speciality, are practitioners of stratigraphy: all geological study is an attempt to unravel piece by piece the history of Planet Earth.

Stratigraphy is also the art of detection. The rocks on the Earth's surface are the clues from which the Earth's history and the processes which have shaped it can be deduced. Each layer or stratum of rock contains a clue to the Earth's geography, climate, and ecology at a specific time. The job of the stratigrapher is to observe, describe and interpret a succession of such layers and other rock bodies in terms of events and processes in the history of the Earth. The rock record which contains the story of our planet can be likened to a book written in a strange language, whose pages have been ripped-out, mixed-up and partially lost. To read the story—a dramatic story of mountain building, climate change and the evolution of life—one must first place the pages of the story in the correct order and then decipher the language upon them. This is the art of stratigraphy.

1.2 The Aim and Structure of this Book

Generations of geologists have come to believe that stratigraphy is boring: simply a recital of geological time periods and of different strata. Like human history, Earth history when taught as a series of facts and dates, can be dull, but it does not have to be. When history is taught in terms of personalities and when both the cause and background to events are discussed, it can be fun, even exciting. The aim of this book is to illustrate that stratigraphy is dynamic and exciting. In order to achieve this the emphasis is placed here on understanding the patterns present within Earth history as opposed to the detailed succession of rock units which make up the stratigraphical record.

The book is divided into two parts. In the first part, we look at the basic information needed to interpret the stratigraphical record by introducing the reader to the 'stratigraphical tool kit' with which the stratigraphy of a region can be built up. Within this kit there are two basic types of tools (Figure 1.1): (1) those tools with

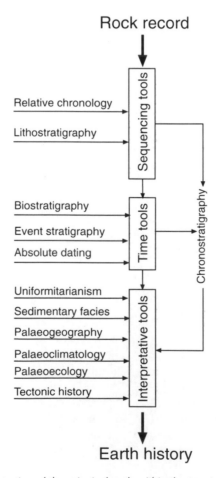

Figure 1.1   *Schematic illustration of the principal tools within the 'stratigraphical tool kit'*

which to establish the succession of rock units, and unravel their relationship in time (sequencing and time tools); and (2) those tools with which to interpret each unit in terms of an event within Earth history (interpretative tools). In the second part of the book, we set out the global pattern present within Earth history: a pattern which has been largely driven by the global concept of plate tectonics. The control of these global events in developing and preserving the stratigraphical record is then discussed and used to structure an account of the present day North Atlantic region, involving both northern Europe and North America, through geological time. A lexicon of key stratigraphical terms and their definitions is given at the rear of the book.

Part One
PRINCIPLES

2

Uniformitarianism or Actualism

In this chapter we introduce the principle of uniformitarianism and explain its importance as an expression of the scientific method adopted in the study of stratigraphy.

2.1 Uniformitarianism

Uniformitarianism is one of the most fundamental geological principles. It can be summarised in the phase 'the present is the key to the past'. In the context of stratigraphy this means that the processes we observe today can be used to unlock the mysteries of Earth history as represented by the record of the rocks.

James Hutton (1726–1797) was the first person to express this principle and grasp its significance for geological time. In 1785 he presented his *Theory of the Earth* which put forward a body of evidence to prove that the hills and mountains of today are being slowly eroded, and that the sediment produced is being deposited as sand and mud on the sea floor to form sedimentary rocks. He argued that the processes of erosion and deposition which can be observed today have always operated. Hutton realised that the vast thickness of sedimentary rocks implied the operation of erosion and sedimentation throughout a period of time, the so-called 'deep time' that he could only describe as being inconceivably long. This contribution was of immense importance at a time when the Earth was widely believed, on the basis of biblical estimates, to have formed at only 4004 BC.

In contrast, Georges Cuvier (1769–1832) interpreted the geological record in terms of a series of catastrophes. In 1796 he was the first person to recognise the presence of extinctions of fossil species within the geological record. To Cuvier, the constant replacement of species through time was clearly a function of major global changes or catastrophes. Cuvier recognised many of these catastrophes within the rock record, but did not invoke supernatural causes. William Buckland (1784–1856), however, interpreted Cuvier's findings in terms of a biblical framework, relating each event to a 'flood', the last of which was believed to be that experienced by Noah.

Charles Lyell (1797–1875) was not disposed towards invoking such catastrophic events in the explanation of Earth history. Lyell was strongly influenced by the ideas of Hutton and was convinced that the world had been shaped by gradual events and that rates of uplift, denudation and of deposition proceeded in a uniform pattern at a constant rate. The term which was coined for this view was 'uniformitarianism'. It was built in part upon Hutton's principle, and was formalised by Lyell in his *Principles of Geology* of 1830. However, Lyell assumed not only the uniformity of process or natural law as Hutton did but also the uniformity of the rate (gradualism). Lyell therefore excluded temporary and local crises, though Hutton's original doctrine did not. During the middle of the nineteenth century there was a continuum between, on the one hand the extreme catastrophists who invoked both present day and unknowable or supernatural causes and on the other, the extreme gradualists, such as Lyell, who allowed only present day processes operating with their present intensity. Lyell's ideas eventually triumphed and dominated the theoretical development of stratigraphy to such an extent that any theory which allowed for the existence of catastrophic influences was deemed unscientific.

In recent years, however, there has been a growing awareness that the stratigraphical record appears to contain, not only sediments produced by slow gradual processes, but also those deposited in more catastrophic environments. This does not mean that the Earth was formed through a series of catastrophes, but reflects the greater probability that catastrophic events are preserved in the stratigraphical record. This can perhaps be best explained in terms of two concepts taken from the study of surface processes. The first of these two concepts is the interrelationship of magnitude and frequency: the greater the magnitude of an event the less frequent its occurrence. Small events may occur several times each year while large events may only occur once in every 100 or 1000 years. For example, hurricanes do not happen every year in southern Britain, but they do occur occasionally, say once every 100 years or so. If you examine the meteorological records of a single year it is therefore unlikely that a hurricane will be recorded, but if you were to examine the records for a period of over 1000 years then the chances of a hurricane being recorded are much greater. In a geological context the longer the time period under consideration, the more likely that the record will contain 'high magnitude' or catastrophic events. We should not, therefore, base our judgement on human time scales. The second relevant concept is that, in general, the bigger the event, the greater its effect and therefore the longer it will take to remove the record left by that event, and the more likely it is that it will be preserved in the sedimentary record. This point is well illustrated in the peat bogs along the east coast of Scotland which contain a record of a catastrophic wave or tsunami: a magnitude of event which appears to have only occurred once in Britain's recent Quaternary record (Box 2.1). Smaller magnitude storm events are not recorded in the sedimentary sequence, but this catastrophic event is.

Box 2.1
A tsunami deposit in eastern Scotland: an example of catastrophism

Within the sequence of Quaternary sediments along the east coast of Scotland there is a thin, regionally extensive sand layer which has been dated as 7000 years old. This sand layer is to be found in the peat bogs and clays (estuarine muds) of the Scottish coast. It extends locally more than 2 kilometres inland from the former coast line and some 4 metres above the former sea level. This sand horizon is believed to have been deposited by a tidal wave (tsunami) centred on the North Sea. The tsunami was probably caused by a small earthquake which triggered a submarine landslide, known as the Storegga slide, in the northern part of the North Sea. The movement of large volumes of sediment and the subsequent sudden changes in sea bed topography can cause dramatic fluctuations in sea level. If the entire volume of the slide failed instantaneously as a block, the theoretical maximum resulting wave may have had a height of 350 metres. The actual wave was probably of much lower height since the failure does not appear to have been instantaneous. This catastrophic event is well recorded in the sedimentary record because the sand layer has been deposited well inland of the area in which conventional marine processes could have eroded it. This catastrophic event is recorded in the sedimentary record, but more conventional events such as storms and tidal surges are not. This example illustrates the way in which the sedimentary record may be biased towards catastrophic events and may not therefore adequately reflect the ordinary geomorphological or depositional processes operating.

Source: Long, D., Smith, D.E. and Dawson, A.G. 1989. A Holocene tsunami deposit in eastern Scotland. *Journal of Quaternary Science* **4**, 61–66.

The result of these observations has, however, led to 'gradualism' now being largely rejected in a geological context. Uniformitarianism expressed in terms of the unity of process – or actualism as it is called in continental Europe – is, however, still the fundamental concept which underlies any attempt to provide a scientific explanation of Earth history. In this context, actualism can be summarised in the statement that 'no powers are to be employed that are not natural to the globe, no action is to be admitted except those which we understand and can observe'.

Given that the aim of stratigraphy is to reconstruct the changing geography, climate and events of the Earth through time there are three main tasks to be tackled: (1) establishing the chronological sequence of rocks; (2) establishing a time history for these rocks; (3) interpreting them in terms of the environment in which they were deposited, that is, in terms of their palaeogeography, palaeoclimate and palaeoecology. Each of these three tasks is dealt with in the following chapters.

2.2   Summary of Key Points

- In the early nineteenth century the geological record was widely interpreted in terms of a series of catastrophic events, of which many were explained in terms of supernatural causes or within a biblical framework.
- Uniformitarianism was developed in the nineteenth century as the antithesis of catastrophism. In the strict Lyellian form, sometimes known as gradualism, it insisted not only upon the uniformity of process but also the uniformity of rates.

- Actualism only assumes the uniformity of process: that is, that the laws of nature have been constant through time. Modern geology recognises that the rates at which processes have operated have varied greatly, and that natural catastrophic events do occur and may be preferentially recorded in the stratigraphical record.
- Actualism is the most fundamental principle in stratigraphy and can be regarded as a statement of scientific method: no powers are to be employed that are not natural to the globe, no process or action is to be admitted except those which we understand and can observe.

2.3   Suggested Reading

Both Dott and Batten (1988) Prothero (1990) and Duff (1993) provide good basic coverage of the subject of uniformitarianism, while the early nineteenth century debate between the theories of catastrophism and uniformitarianism is well-presented by Hallam (1983). Gould (1988) provides an interesting account of the discovery of time in stratigraphy, and introduces the concept of 'deep time'.

References

Dott, R.H. and Batten, R.L. 1988. *Evolution of the Earth*. McGraw-Hill, New York.
Duff, P. McL. D. (Ed.) 1993. *Holmes' Principles of Physical Geology*. Chapman Hall, London.
Gould, S.J. 1988. *Time's Arrow, Time's Cycle*. Penguin Books, Harmondsworth.
Hallam, A. 1990. *Great Geological Controversies*. Oxford University Press, Oxford.
Prothero, D.R. 1990. *Interpreting the Stratigraphic Record*. Freeman and Co., New York.

3

Establishing the Sequence of Events

Stratigraphy is about interpreting the Earth's geological record as a sequence of events through time. As each sequence of rocks is equivalent to a sequence of events, it is necessary to have the tools to interpret and understand the relative order of these events. In this chapter we explain the two basic principles of relative ordering, superposition and relative chronology and then show how they can be applied to establish the sequence in the field.

3.1 The Principle of Superposition

The principle of superposition states that in any undisturbed sedimentary sequence, the layer at the bottom of that sequence was formed first. Superposition is therefore the most fundamental principle in determining the relative order of rock units in a sequence.

The principle was developed in the seventeenth century, when Niels Stensen (1638–1686), a Danish physician living in Florence, was invited to dissect a large shark. He was careful to observe and describe the shark in minute detail. In the course of his work, Stensen, often known by his anglicised name of Steno, recognised that the teeth of the shark closely resembled the stony fossils, known as Glossopetrae, found in the uplands of Tuscany. At this time the idea that fossils were the preserved remnants of past organisms was not widely held, but Stensen by careful observation demonstrated that the Glossopetrae and the shark's teeth were identical in form (Figure 3.1).

This discovery demonstrated that fossils were actually the remains of once-living animals, and that the rocks of Tuscany were formed in a sea which had once covered the land. He recognised that fossils were actually incorporated in rocks before they

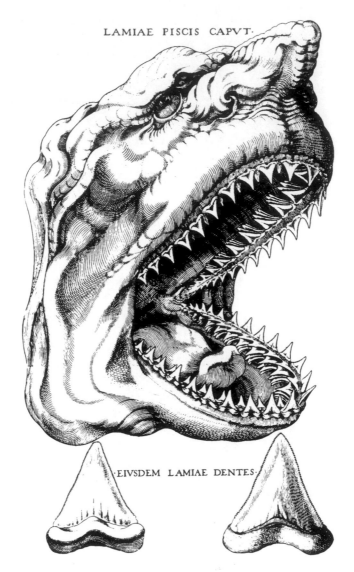

LAMIAE PISCIS CAPVT·

·EIVSDEM LAMIAE DENTES·

Figure 3.1 *Steno's comparison of shark's teeth with Glossopetrae. The living shark was dissected by Steno and the teeth compared with the fossil objects known as Glossopetrae, illustrated beneath the shark's head. This illustrated clearly for the first time the organic nature of fossils. [From: Mercati (1717) Metallotheca, The Vatican, p. 333. Reproduced with the permission of the Natural History Museum, London]*

were lithified, and that the deposition of sediments in the sea ultimately led to the formation of fossil-bearing rocks in layers. The outcome of Stensen's investigations was the development of the principle of superposition.

Stensen published the results of his work in 1669 and his results can be summarised in three statements:

1. Superposition: when a layer of rock was forming, the one below had already formed and was therefore older.
2. Original horizontality: layers of rock were originally deposited horizontally.
3. Original lateral continuity: layers of rocks extend laterally until physically constrained in some way.

Using these three statements, Steno was able to propose a relative chronology for the layers of strata within the Tuscany area. A chronology is a list of events in the order in which they occurred.

3.2   Superposition

The principle of superposition allows us to identify this order in any succession of undisturbed sedimentary rocks which are stacked in layers one upon another; the oldest will always be found at the bottom (Figure 3.2). Determining the chronology of rock units in a vertical sequence involves establishing the relative time relationships between each unit of the sequence. The discovery of superposition provides the mechanism by which bedded rocks can be recognised as the products of a chronological sequence of depositional events. Today, most geologists take this principle for granted, but its importance cannot be underestimated. From it, geologists have a

Figure 3.2   *Succession of horizontally bedded strata (Blue Lias) from the Lower Jurassic of Lyme Regis, Dorset, England. [Photograph: BGS]*

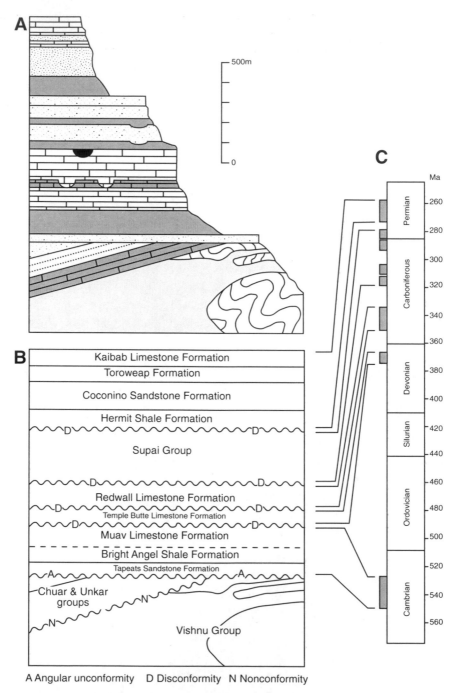

Figure 3.3   Stratigraphy of the Grand Canyon. The Grand Canyon exposes a series of horizontally bedded strata overlying much older, folded and metamorphosed rocks. Application of the principle of superposition allows us to construct a stratigraphical column. [Modified from: Press & Siever (1986) Earth, Freeman, Figure 2.14, p. 32]

simple method by which they can determine the relative order of undisturbed strata. The Grand Canyon in Arizona provides a good example of the principle of superposition in action (Figure 3.3). Here an unusually complete geological section through the Earth's sedimentary rock record can be seen in the sides of the canyon cut by the Colorado River. At the base there are ancient rocks which underlie a sequence of sedimentary layers, stacked one on top of another. Application of the principle of superposition leads us to conclude that the oldest layer of rock is at the base and that each successive layer decreases in age. The sequence of rock layers can be summarised as a stratigraphical column, in which, by convention, the oldest rocks are presented at the base of the column, thereby mimicking nature (Figure 3.3).

Even in deformed sequences, where the layers have been folded and the definition of the original top and bottom of the sequence is ambiguous, superposition can be applied with the aid of what are known as right way-up structures. These structures provide a record of the original way-up of the sedimentary sequence by defining the original top and bottom of the sequence. Examples of right way-up structures are given in Figures 3.4 and 3.5, while their application in the field is illustrated in Box 3.1.

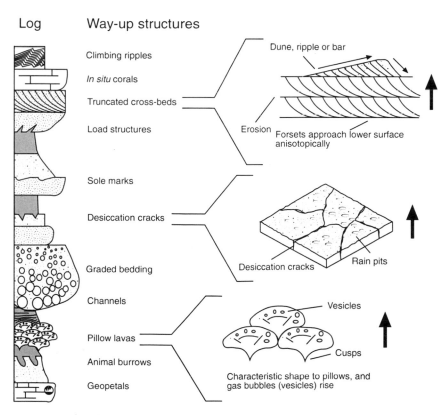

Figure 3.4   *Examples of some common way-up structures*

Figure 3.5 *Photographs of typical way-up structures.* **A**: *Graded bedding in the Torridonian Sandstone of north-west Scotland. Within each unit of graded bedding the particles are coarse at the bottom and become finer upwards [Photograph: BGS].* **B**: *Recent sedimentary features formed on a muddy puddle in the floor of a quarry by the action of rain (rain pits) and sun (dessication cracks). Both these structures may be preserved in the sedimentary record and can be used as way-up criteria [Photograph: BGS]*

Box 3.1
Way-up criteria in stratigraphical studies

In 1958 R.M. Shackleton was first to demonstrate that the rocks in the Aberfoyle anticline were actually upside down. This anticline forms part of an area of the highland borders in Scotland that has been intensely deformed during the Grampian Orogeny. Although the rock sequences involved have been heavily affected by the tectonic activity, the sequence of turbidite sandstone known as the Aberfoyle Grits still retain their primary sedimentary structures, and particularly, their graded bedding. Shackleton demonstrated that this graded bedding was consistently reversed in Aberfoyle, and from this he concluded that the structure was actually inverted and formed part of the much larger Loch Tay belt – a large, complexly overfolded structure associated with the Grampian Orogeny. This example demonstrates the value of detailed stratigraphical studies to the understanding of the development of a complex part of the Earth's crust.

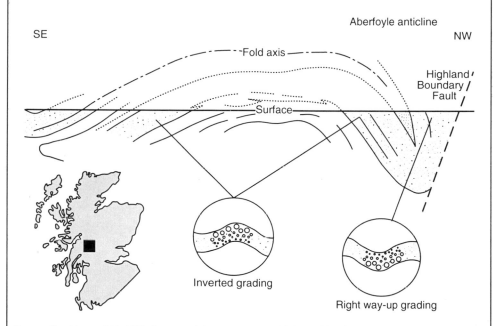

Source: Shackleton, R.M. 1958. Downward-facing structures of the Highland border. *Quarterly Journal of the Geological Society of London* **58**, 361–392.

3.3   Determining Relative Chronology

On its own the principle of superposition cannot resolve the chronology of complex geological areas, in which the rocks are, for example, not simply in horizontal layers. In such areas geologists require more sophisticated tools with which to establish the relative chronology of both rock units and geological events. The most important of these tools are cross-cutting relationships and unconformities. Before looking, however, at each of these tools in detail we shall first look at the development of these ideas by early geologists.

Superposition as a tool was applied literally by many early geologists. Geologists such as Abraham Gottlob Werner (1749–1817) argued that geological strata continued infinitely around the globe and were precipitated from a primeval ocean which once covered the globe. Each layer was horizontal and indicative of Stensen's principle of superposition. Werner divided the world's geological succession into four units:

1. The Primitive or crystalline rocks which were precipitated from this primeval and universal ocean.
2. The Transition rocks containing fossils indicating the creation of life.
3. The Flötz deposited in running water formed by the contraction of this universal ocean.
4. The Alluvial rocks formed by the flow of water on land.

Werner interpreted all rocks, including crystalline granites and basalts, as being formed by sedimentation in an ocean or by running water. Implicit in his model of the geological record was the fact that deformation of rock layers and surface erosion had only occurred once within Earth history (during the Flötz and Alluvial periods). This model of Earth history was consistent with the ideas of the theologians of the time who considered the Earth to have been formed just 6000 years ago.

The first person to really appreciate that Werner's scheme was inadequate to explain the development of the Earth was James Hutton. Hutton approached the interpretation of the geological record on the basis of sound field observations backed by deductive reasoning. His researches led him to question the ideas of Werner and his contemporaries. He made two important observations.

First, at Glen Tilt, in the Cairngorm Mountains of Scotland, Hutton noted that granite appeared to penetrate the surrounding metamorphic schists in a series of veins (Figure 3.6). He concluded that the granite must have been in a molten state for it to have forced its way into the schist in this way. Hutton made two inferences from this observation: (1) granite is not precipitated from water as suggested by Werner and his contemporaries, but is formed by the cooling of molten rock; and (2) that the granite must be younger than the schists into which it was forced as a fluid mass. The second of these inferences is an example of a cross-cutting relationship and is important in determining the relative chronology of geological events as we shall see in the next section.

Second, at Siccar Point in Berwickshire, southern Scotland, and at other localities on the coasts of Scotland and northern England, Hutton made a discovery that illustrated that the age of the Earth was much in excess of that estimated on the basis of the biblical account of creation. He suggested that periods of rock deformation, uplift and erosion had occurred more than once in the geological record. At Siccar Point, Hutton observed two bodies of strata apparently at right angles to each other. The older shale strata were arranged so that the beds were almost vertical, while the younger overlying sandstone strata lay almost horizontally on top of the vertical beds (Figure 3.7). In view of the fact that the two units did not conform to the normal pattern of one horizontal layer on top of another the relationship was termed an unconformity. Importantly, eroded fragments of the older strata were

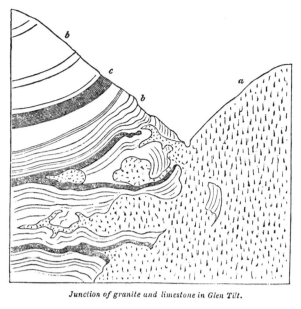

Junction of granite and limestone in Glen Tilt.

a, Granite.   *b,* Limestone.   *c,* Blue argillaceous schist.

Figure 3.6 *Lyell's (1830) illustration of Hutton's famous locality at Glen Tilt, Cairngorm Mountains, showing the true nature of granite. Hutton appreciated from this site that the granite must have been intruded in a molten state into the schists and limestones of the surrounding area. This cross-cutting relationship clearly illustrates the relative age of the granite, which is younger than the surrounding schist rocks. [From: Lyell (1830)* Principles of Geology, *John Murray, Figure 88. Reproduced by permission of the Natural History Museum, London]*

found in the lower bed of the topmost formation. If sedimentary rocks were laid down originally on a horizontal plane, as Stensen had taught, then the underlying sequence must have undergone a cycle of deposition, tilting and erosion, before the overlying rocks, which incorporated blocks of the older strata, were then deposited on top. As John Playfair (1742–1819), one of Hutton's contemporaries, put it in 1802, 'what clearer evidence could we have had of the different formation of these rocks and of the long interval that separated their formation'. Hutton drew two inferences from these observations: (1) that there had been several periods of deformation, uplift and erosion, not one as suggested by Werner; and (2) the Earth must be of much greater antiquity than the predictions made from the Bible to allow sufficient time for this to occur.

Cross-cutting relationships and unconformities are therefore very important in determining the relative chronology of geological strata and are discussed further in the following sections.

3.3.1 Cross-cutting Relationships

Hutton's discovery at Glen Tilt established the principle that rock sequences are older than the geological features or structures, such as faults or igneous intrusions, which

Figure 3.7 *Photograph of Hutton's unconformity at Siccar Point, Berwickshire, southern Scotland. Unconformity of Upper Old Red Sandstone (Devonian) on the vertical Silurian greywackes and shales [Photograph: BGS]*

cut across them. The principle of cross-cutting relationships is particularly important in determining the chronology of rock units where other independent data are unavailable. Figure 3.8 shows a typical example of a geological sequence in which cross-cutting relationships are present. In this example the sedimentary sequence of layers 1 to 6 were deposited first. The dyke (D) was intruded after the sedimentary

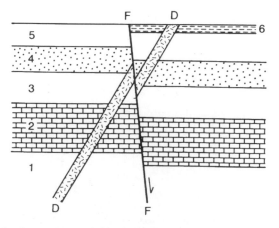

Figure 3.8 *An example of cross-cutting field relationships, see text for explanation*

succession was deposited. This chronology is apparent because of the way that the igneous body cuts across the sedimentary succession. Within larger igneous bodies, fragments of the surrounding sedimentary succession may be incorporated into the molten rock as it is forced in position. These fragments, known as xenoliths, help confirm that the surrounding succession, the country rocks, were formed first. Continuing with our example, a fault (F) subsequently developed and this displaced, not only the original sedimentary layers, but also the later igneous dyke (Figure 3.8). Therefore the principle of cross-cutting relationships allows us to recognise three events in our example (Figure 3.8) with the following relative chronology: (1) deposition of beds 1 to 6; (2) intrusion of the dyke, and; (3) formation of the fault and displacement along it.

The principle of cross-cutting relationships is extremely valuable as a methodological tool for establishing relative chronology and is particularly useful in rock successions which have been subsequently folded, faulted and intruded by igneous rock. A detailed example of the application of this tool in the geological record is presented in Section 7.1.1 of this book.

3.3.2   Unconformities

Unconformities allow the recognition of a relative chronology of events, but also indicate the presence of time gaps within rock sequences. These gaps may span very long or very short intervals of time. Four types of unconformity can be recognised in the stratigraphical record, each formed under different circumstances. The four types are (Figure 3.9): (1) Angular unconformity, (2) Disconformity, (3) Nonconformity, and (4) Paraconformity.

Hutton's original unconformity at Siccar Point (Figure 3.7) displays a marked angular discordance between the rock layers beneath and above the plane of unconformity, and provides therefore a good example of an angular unconformity. Put crudely, the rock units above and below the plane of unconformity have different angles of dip. The history of this unconformity involves the following sequence of processes: deposition → deformation → uplift → erosion → re-deposition. This cycle of events is illustrated in Figure 3.10 and can be broken down into four stages. First, a sequence of sedimentary rocks was deposited horizontally, according to Stensen's principle of superposition. Second, this sequence was then folded and uplifted, probably as a result of mountain building caused by continental collision (see Section 6.6). Third, the uplifted and folded rocks were subject to erosion and the landscape was dissected and lowered. Finally, a rise in sea level flooded this eroded landscape and a new sequence of sedimentary rocks was deposited horizontally above the older deformed sequence.

The three other types of unconformity are more subtle. A disconformity occurs where the unit above and below the plane of unconformity have the same angle of dip, but are separated by a surface which has undergone erosion (Figure 3.9). Here, the older sequence of rocks has not been folded, but has been subject to erosion before the deposition of the younger rocks above the unconformity. This type of unconformity may occur because of a fall in sea level which exposes the rocks of the sea floor to

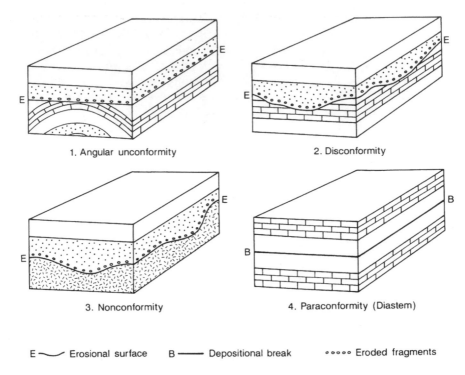

Figure 3.9   *The four main types of unconformity. [Modified from: Dunbar & Rodgers (1957), Principles of Stratigraphy, John Wiley & Sons, Figure 57, p. 117]*

subaerial erosion, before a subsequent sea level rise leads to the deposition of new sedimentary rocks on top of the erosional surface. The erosion surface represents a break or time gap in sedimentation. A classic example of a disconformity is seen in the Hampshire and London basins of southern England. Here an eroded and potholed Cretaceous Chalk surface denotes a major erosional break in the sequence between the Cretaceous Chalk and the younger Eocene sediments above (Figure 3.11).

Where bedded sediments overlie an exposed and eroded crystalline basement, the term nonconformity is used (Figure 3.9). Often this is indicative of a landscape formed by the erosion of intensely deformed and metamorphosed rocks produced during continental collision, although it can be produced when a large igneous intrusion is exposed by erosion. Deposition of horizontally layered sedimentary rocks, during a rise in sea level, or by river systems, produced the nonconformity. In Britain a good example of this is the deposition of the Torridonian red sandstones of the Late Proterozoic on top of the much older, dissected ancient land surface of the crystalline Lewisian basement, in north-west Scotland. All the different types of unconformity discussed so far are visible in the succession of rocks exposed in the sides of the Grand Canyon as shown in Figure 3.3.

Linking all of the unconformities discussed is the fact that each is characterised by evidence of erosion during which time there was no deposition within the area and consequently the erosional surface represents a time break within the sedimentary

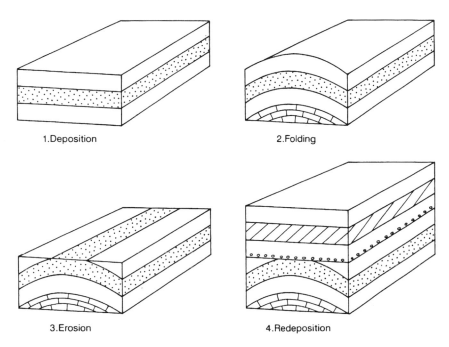

1.Deposition   2.Folding

3.Erosion   4.Redeposition

Figure 3.10 *The cycle of events which leads to an angular unconformity. The diagram shows the four phases in the production of an angular unconformity: (1) the deposition of horizontally bedded sediments; (2) folding and uplift of the sedimentary strata; (3) the erosion of the folded strata; (4) the deposition of new sedimentary rocks on top of the eroded surface.*

sequence. In many cases, fragments may be eroded out of the underlying sequence and reincorporated or reworked into the covering rocks. However, we can also recognise an unconformity, or time break, where there has been little or no erosion, but where the net rate of sedimentation has matched the net rate of sediment loss and consequently there is no sedimentary record of the time period in question. This type of unconformity is known as a paraconformity. In these cases the sediment sequence is not exposed to subaerial erosion, and recognition of the existence of the paraconformity is carried out purely on the absence of time-significant fossils, or indeed, on the accumulation of such fossils with very little background sediments in condensed deposits, discussed more fully below.

The gaps produced in the sedimentary record by non-deposition can be of many different orders of magnitude, representing a few hours or days, years or even millions of years. The length of time represented by such gaps can only be determined by recourse to an independent means of dating. Some geologists believe that most stratigraphical successions contain numerous diastems, obscure or subtle unconformities of uncertain origin (Box 3.2). In fact it has been argued that the stratigraphical record is rather like a net – a series of holes held together by string – and that each bedding plane is in fact a period of time when there was no deposition and is therefore a gap in the geological record. In practice the extent to which the record is complete or incomplete is uncertain, open to debate and likely to vary from one depositional

Figure 3.11   *Photograph of the disconformity at the base of the Palaeogene deposits of the London Basin. The photograph illustrates the uneven surface of the eroded Cretaceous chalk strata at Pegwell Bay, Kent, England. This surface is overlain by younger, bedded sands, deposited during the Palaeogene [Photograph: D.G. Helm]*

Box 3.2
Episodic deposition and preservation: more gaps than record?

The idea that the rock record is composed mainly of gaps is illustrated by the work of Loope (1985) on the aeolian sandstones of the Late Palaeozoic in south-eastern Utah. In the Canyonlands National Park, Utah, the 200 metre thick Cedar Mesa Formation consists of units of cross-bedded sandstone formed by the climbing aeolian dunes, separated by prominent flat-lying bedding planes. Detailed sedimentological investigation of these bedding surfaces suggests that they represent disconformities; surfaces subject to erosion and subsequent colonisation by vegetation. The sandstones appear to have been deposited episodically. Periods of rapid sand accretion, associated with episodes of climbing dunes, were punctuated by episodes of deflation. The episodes of dune accretion were probably triggered by sea level regression of adjacent marine environments, which provided an upwind sand supply. Each of the bedding surfaces within the Ceda Mesa Formation represents a time break or diastem. This example illustrates how a rock record can be built up by episodes of rapid sedimentation separated by extended periods of non-deposition; a classic example of more gaps than record.

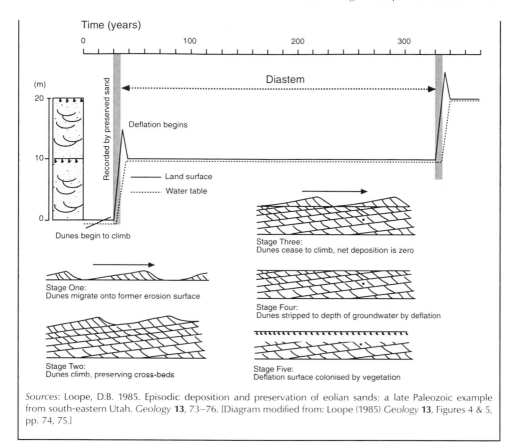

Time (years)

Diastem

(m)

Recorded by preserved sand

Deflation begins

Land surface

Water table

Dunes begin to climb

Stage One:
Dunes migrate onto former erosion surface

Stage Two:
Dunes climb, preserving cross-beds

Stage Three:
Dunes cease to climb, net deposition is zero

Stage Four:
Dunes stripped to depth of groundwater by deflation

Stage Five:
Deflation surface colonised by vegetation

Sources: Loope, D.B. 1985. Episodic deposition and preservation of eolian sands: a late Paleozoic example from south-eastern Utah. *Geology* **13**, 73–76. [Diagram modified from: Loope (1985) *Geology* **13**, Figures 4 & 5, pp. 74, 75.]

environment to the next. What is clear, however, is that the record is sufficiently complete in most cases to allow a reading of Earth history to be made, and this is the practical reality of the day-to-day operation of stratigraphical studies.

3.4   Establishing the Sequence in the Field

The first task of any geologist entering a new field area is to establish the sequence of rocks present. This involves three things: (1) detailed observation and description of the rocks (lithologies) present; (2) formal establishment of stratigraphical units according to convention and; (3) the determination of their distribution in space. This process is the subject of lithostratigraphy.

Lithostratigraphy is the formal description of rock units and their comparision with others in both space and time. It provides much of the raw data needed to interpret the geological record. It is fundamental to the process of comparing the stratigraphical record of one area with another in order to build up a picture of a region's geology. It is also the first stage in determining the time relationship of rocks in one area with those of another distant area.

Lithostratigraphy entails the division of rock sequences into units, known as lithostratigraphical units, on the basis of lithology (rock type). Each of these units should be lithologically homogeneous and their recognition is helped when they have distinct lithological boundaries. Traditionally, lithostratigraphical units have been defined for only layered or stratified rocks. The geological record is, however, made up of more than just sedimentary rocks and lithostratigraphy should encompass the description of all rock units, including igneous and metamorphic rocks. The term lithodemic unit is used by some authors when referring to rock units which are not stratified. This term has not, however, found universal favour, although it is reasonably well accepted in North America.

Lithostratigraphy employs a strict hierarchy of terms for units so that geologists worldwide understand what is meant by each term (Table 3.1). This terminology is governed by international agreement. The fundamental unit is the formation, which is described as a unit of largely homogeneous lithology that may be clearly recorded on a geological map. A formation is therefore a mappable unit. It is only established formally when it has been described and published in the geological literature. This ensures that different geologists use the same name for the same unit, which prevents confusion and misinterpretation. This formal procedure is described in Box 3.3. Description of these mappable units forms the basis for lithostratigraphy.

Lithostratigraphical units are said to crop out at the Earth's surface and it is usual to refer to their lateral extent at the surface as their outcrop. The distribution of lithostratigraphical units is best recorded on an outcrop map (i.e. a geological map). The construction of such maps is an important part of the stratigrapher's task and one which is made difficult by the fact that the outcrop of a geological unit is often covered by soils, vegetation or buildings. Geological maps are therefore usually made by study of, and extrapolation from, the rocks which are actually visible, or exposed, at the surface.

The comparison of lithostratigraphical units with others in different locations is known as correlation. Usually comparison is made with sequences in adjacent areas, but correlation may be achieved between widely spaced localities, for example between continents. The ultimate aims of correlation are: (1) to establish the relative

Table 3.1   *The hierarchy of formal lithostratigraphical units*

Lithostratigraphical Unit	Definition
Supergroup	Clustering of groups based on their lithological characteristics or on their mode of formation
Group	Clustering of formations based on their lithological characteristics or on their mode of formation
Formation	Mappable unit of homogeneous lithology
Member	Subdivision of a formation
Bed	Lithologically distinct horizon or layer

Box 3.3
Formal lithostratigraphy in practice

The practical principles of lithostratigraphy are well illustrated by the work of Ineson et al (1986) on James Ross Island in Antarctica. This is pioneering work in an area of little geological exploration, and as a consequence one of the first tasks was to establish a formal lithostratigraphy which the paper by Ineson et al (1986) does. In their study Ineson and his colleagues first reviewed previous work, before establishing their own lithostratigraphy on the basis of rigorous and detailed field observations. The rocks exposed on the western side of James Ross Island are of Cretaceous age. Ineson et al (1986) recognise two groups – Marambio Group and Gustav Group – and seven formations. As illustrated in the geological sketch maps below each formation is a mappable unit, with an internally homogeneous lithology and distinct boundaries. Each of the groups and formations was formally defined with reference to five parameters dealt with systematically in the paper: (1) the lithology was described; (2) type or reference sections were identified and located on maps of the area; (3) formal names were chosen on the basis of a place, a lithology and a unit rank (e.g. Whisky Bay Formation); (4) the nature of mappable boundaries was discussed; and (5) the thickness of each unit was recorded. The work of Ineson et al (1986) provides an exemplar of how a formal lithostratigraphy should be established and reported in the literature for future use.

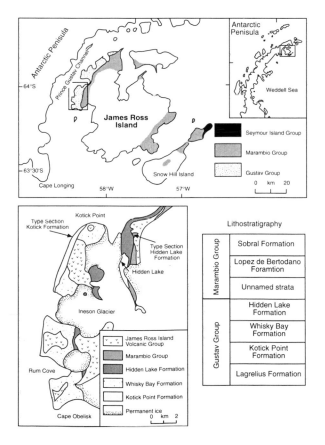

Source: Ineson, J.R., Crame, J.A. and Thomson, M.R.A. (1986). Lithostratigraphy of the Cretaceous Strata of West James Ross Island, Antarctica. *Cretaceous Research* 7, 141–159.

chronology of lithostratigraphical units and therefore the relative sequence of geological events over a wide area, and (2) to provide an understanding of the geological development of a specific area through a knowledge of the spatial pattern of rock units within it. In practical terms such correlation may be achieved by a variety of means. Simple visual matching of lithological succession is the commonest method used. Lithological equivalency is established by drawing tie lines (Figure 3.12) to link lithological units which have a similar composition. Tie lines make no assumption of time equivalency; the same rock type may form in two different areas at different times. In contrast, time lines are independent of lithology and are established using time-significant tools such as fossils or sudden geological events; these tools are discussed in Chapter 4. Tie lines may cross time lines as illustrated in Figure 3.12.

Conventional lithostratigraphy is sometimes limited by the availability of geological exposure; if you can not see the different lithologies, you can not define lithological units. Where exposure is limited lithological correlation may be achieved in some cases by the application of geophysical methods in boreholes (wells). These methods involve the description and subsequent correlation of rock units on the basis of their geophysical properties. Over the last 30 years the sedimentary basins of the world have been probed by numerous wells in the quest for oil and gas. Information is obtained both during the drilling of a well and from downhole observations within it. For example, the speed at which a well is sunk provides information on the relative

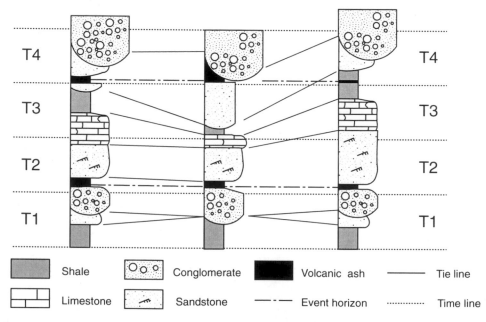

Figure 3.12 Physical correlation and the relationship of tie lines and time lines. The three sedimentary successions illustrated can be correlated on the basis of lithological similarity. The lithological units are linked with tie lines. Time lines link points of equal age. They may be constructed on the basis of faunal evidence, in which case they are independent of lithology and time lines may cross tie lines. Time lines may also be based on lithological event horizons, or isochronous surfaces (e.g. layers of volcanic ash)

hardness of the rocks, which may also be sampled or examined from the drill chips produced. Once a well has been sunk a measuring device, known as a sonde, is lowered on a cable or wire. A range of geophysical observations are made by the sonde as it is lowered and the results are plotted against depth to give a wire-line or well-log. Table 3.2 lists some of the common properties that are measured and the sort of information that can be obtained. Correlations between well-logs are made in a range of different ways. For example, distinctive peaks in the record of one log can be correlated with peaks in another log, or lithological signatures (i.e. groups of peaks representing sections of log) can be correlated between logs. In locations where well-logs have been used extensively, such as in the North Sea, type wells are sometimes defined to represent lithological units (Figure 3.13); these are then used as the basis for correlating well-logs and ultimately the lithological units they contain.

Table 3.2   *Main types of geophysical observation in wire-line logging*

Name	Measures	Typical uses and applications
Caliper Logs	Measures borehole diameter, size and shape	Information on rock properties, such as lithology, hardness and permeability
Resistivity Logs	Electrical currents are passed through the well walls and current resistance measured	Most rocks are insulators, so current flow is through pore water and therefore it provides information on porosity and pore fluid content
Dipmeter	Microresistivity measurements are made simultaneously on different faces of the well wall and correlated to give data on dip	Information on the angle of dip of beds and horizons
Spontaneous Potential Logs	Measures the electrical currents that occur naturally in boreholes as a result of salinity differences	Information on the presence of permeable beds
Gamma Ray and Spectral Gamma Logs	Measures the natural radioactivity of the rock in the borehole wall produced by uranium, thorium and potassium. Gamma Ray Logs measure the total radioactivity while Spectral Gamma Logs differentiate between the three main sources	Identification of radioactive minerals, often concentrated at unconformities. Shales have a very high gamma radiation output, so information can be obtained about grainsize variations and therefore depositional environments (facies)
Neutron Logs	Neutrons are emitted from a radioactive source on the sonde; loss of energy is measured. Most energy is lost when a neutron collides with a hydrogen atom.	Data on the amount of liquid-filled porosity
Density Logs	The rock is bombarded with gamma rays and the amount of energy loss provides information on bulk density	Information on porosity, mineral content, lithology and detection of gas
Sonic or Acoustic Logs	Measures the speed of sound through a rock in the well wall	Information on lithology, porosity, compaction and burial history

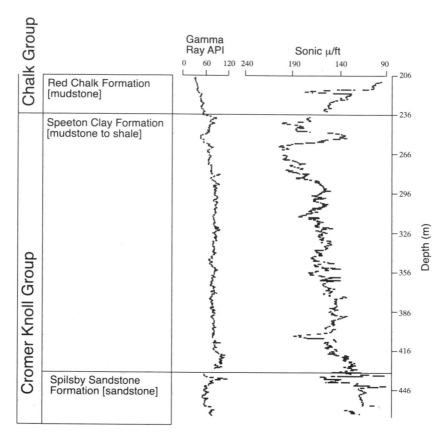

Figure 3.13   *Geophysical signature for a Cretaceous succession in the North Sea. [Modified from: Rider, M.H. (1986), The Geological Interpretation of Well Logs, Whittles Publishing, Figure 13.1, p. 155]*

In order to understand the distribution, form and significance of lithological units it is important to appreciate three important concepts: (1) distribution in space; (2) distribution in time; and (3) relative thickness.

The distribution in space or continuity of a single unit will depend on the continuity of the original sedimentary environment in which it was deposited. For example, the sediments deposited in marine environments are often much more uniform in their lateral extent than those deposited in a fluviatile (river) environment, the boundaries of which are determined by those of the river valley. This is discussed further in Section 5.1.

As first demonstrated by Stensen in the seventeenth century, most sedimentary rocks are deposited in horizontal layers. However, such layers may accumulate in one of two ways: (1) by vertical accumulation, due to a gentle rain of sediments from above; or (2) by lateral accumulation from specific points, such as the mouth of a river, from which sediments build out. Most contemporary sediments, away from deep ocean basins, accumulate laterally. Within the geological record sedimentary layers, which accumulated laterally, may be indistinguishable from those which

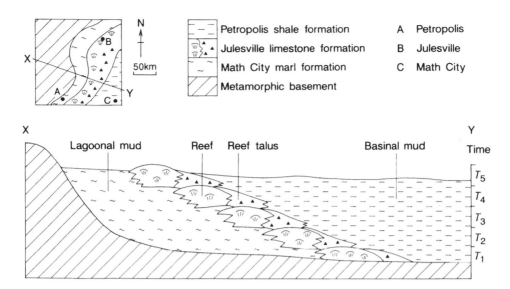

Figure 3.14   *The concept of diachronism. With rising sea level the facies shift landwards through the time interval from T_1 to T_5. This process leads to the creation of rock units of uniform lithology. These may be mapped and named as formations, although each may have been formed at different times through the extent of the outcrop. [Reproduced with permission from: Doyle (1996), Understanding Fossils, John Wiley & Sons, Figure 5.2, p. 96]*

accumulated vertically. This has implications for the distribution in time of lithological units. Units which have accumulated laterally in space will have formed at progressively different times in different places. This can be illustrated with reference to Figure 3.14, which shows a section through a barrier reef. As sea level has risen over the five time intervals shown (T1 to T5) the reef has moved successively shoreward. Ultimately the units of the reef deposits, lagoonal muds, and offshore muds can be correlated on the basis of lithological similarities, but the lithological boundaries which link the units are not time significant.

Lithostratigraphical units which accumulate in this way are not therefore equivalent to time units and are said to be diachronous – that is, although the unit is uniform in lithology, it has formed at different times in different places. Another problem exists when considering the distribution of lithostratigraphical units in time. Given the range of depositional environments present today, it follows that we would expect a similar range of environments, with similar products, to have recurred through geological time. For example, desert sandstones have formed more than once in Earth history and their occurrence is not therefore time significant. As a consequence we should not assume that because two rock units are of the same lithological type, they were formed at the same time. In other words, our examples demonstrate that lithostratigraphical units are not usually in themselves time significant; in order to compare units of the same age an independent means of correlation is needed. The most common tools for correlation are fossils, and their use as stratigraphical tools is known as biostratigraphy, a technique which is discussed in Section 4.1.2.

Box 3.4
Basins and swells in the Jurassic

The Lower and Middle Jurassic rocks of England were deposited over a broad area of flooded continental shelf, and comprise a range of mudrocks and carbonates. Although the topography of this flooded area is thought to have been low, there are some parts which experienced greater amounts of subsidence than others. The areas experiencing the least amount of subsidence, known as swells, seem to equate with regions that were either buoyed-up by igneous intrusions (e.g. Market Weighton) or had underlying structural anomalies. The observed result in these positive areas is that stratigraphical sequences are relatively condensed in comparison with the adjacent subsiding basins and their expanded sequences. The accompanying figure illustrates the concept. The thickness of individual biozones provides a way of directly comparing the thickness of sequences in specific areas. In the Market Weighton area (locality 2) and the Mendip Hills (locality 4), for example, the topography was relatively high and the Jurassic rocks are condensed, being largely thin or absent. In contrast, in the adjacent Yorkshire (locality 1) and Cotswold (locality 3) basins there is a greater thickness of sediment preserved, representing an expanded sequence.

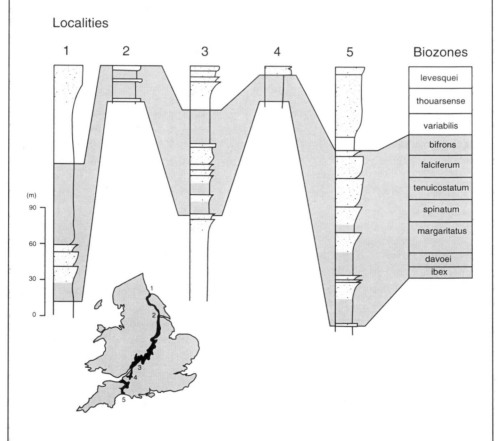

Source: Sellwood, B.W. and Jenkyns, H.C. 1975. Basins and swells and the evolution of an epeiric sea (Pliensbachian–Bajocian of Great Britain). *Journal of the Geological Society of London*, **131**, 373–388.
[Modified from: Sellwood and Jenkyns (1975) *Journal of the Geological Society of London* **131**, Figure 1, p. 374.]

The relative thickness of a deposit or unit may also be misleading. There are many instances of immensely thick deposits in one area having been deposited over the same time interval as deposits of only a few metres thickness elsewhere. For example, the Lower Jurassic Junction Bed, exposed on the south coast of England in Dorset, is only one metre thick but is the exact time equivalent of the Whitby Mudstone Formation which is over 50 metres thick and exposed in north-east England. This difference reflects deposition in two different types of sedimentary basin – a subsiding marine basin in the case of the Whitby Mudstone Formation (Yorkshire), and a submerged topographically high area in the case of the Junction Bed (Dorset). Thin successions like the Junction Bed, which have much thicker time equivalent deposits elsewhere, are said to be condensed (Box 3.4), and may result from many diastems. In summary, no assumption of time should be made from the thickness of a unit.

3.5   Summary of Key Points

- The principle of superposition, which states that in an undisturbed sequence the lowest sedimentary layer is the oldest, allows the first recognition of the relative stratigraphical order of a series of rock units. The application of way-up criteria ensures the order of superposition can be accurately determined even in deformed and inverted sequences.
- The principle of relative chronology enables a more detailed breakdown of stratigraphical sequences. Cross-cutting relationships are particularly important in metamorphic and igneous terrains. Unconformities help to establish sequences of events in geological time, and record the presence of time gaps.
- The first task in stratigraphy is the recognition, description and correlation of lithological units. This is known as lithostratigraphy.
- Many lithological units accumulated laterally and are therefore fundamentally diachronous: they are lithologically, but not temporally identical throughout their extent.
- The relative thickness of correlated units is not necessarily indicative of the time taken in deposition. Immensely thick units may be deposited rapidly: relatively thin units may have accumulated slowly.
- Correlation between rock units may be achieved through matching lithologies, or through comparing the geophysical properties of individual lithologies. The lines of correlation joining lithologies are known as tie lines. Tie lines are independent of time relationships, which can only be determined using independent, time specific tools, such as fossils or geological events.

3.6   Suggested Reading

Ager (1993) is a thought-provoking and highly influential reader which should form the basis for further discussion of the nature of the stratigraphical record. These ideas are also examined by Dott (1983). Dott and Batten (1988) and Duff (1992) provide further introduction to some of the concepts introduced in this chapter, while Boggs (1987) explores these concepts in more detail. Hedberg (1976) provides a detailed guide to stratigraphical procedure as recommended by the International Union of Geological Science, while Whittaker et al (1991) provide an abbreviated version of this subject. Barnes (1988) gives a detailed guide to the description of rock units in the field. Cox and Sumbler (1998) provide an advanced review of the principles and practice of lithostratigraphy, while Whittaker (1998) gives a useful introduction to geophysical well-logs.

References

Ager, D.V. 1993. *The Nature of the Stratigraphical Record*, (Third Edition). John Wiley, Chichester.

Barnes, J. 1988. *Basic Geological Mapping*. Geological Society of London Handbook, Open University Press, Milton Keynes.

Boggs, S. 1987. *Principles of Sedimentology and Stratigraphy*. Merrill, New York.

Cox, B.M. and Sumbler, M.G. 1998. Lithostratigraphy: principles and practice. In Doyle, P. and Bennett, M.R. (Eds) *Unlocking the Stratigraphical Record*. John Wiley, Chichester, 11–28.

Dott, R.H. 1983. Episodic sedimentation – How normal is average? How rare is rare? Does it matter? *Journal of Sedimentary Petrology* **53**, 5–23.

Dott, R.H. and Batten, R.L. 1988. *The Evolution of the Earth*, (Fourth Edition). McGraw Hill, New York.

Duff, P. McL. D. (Ed.) 1993. *Holmes' Physical Geology*. Chapman & Hall, London.

Hedberg, H.D. (Ed.) 1976. *International Stratigraphic Guide: a Guide to Stratigraphical Classification, Terminology and Procedure*. John Wiley, New York.

Whittaker, A. 1998. Borehole data and geophysical log stratigraphy. In Doyle, P. and Bennett, M.R. (Eds) *Unlocking the Stratigraphical Record*. John Wiley, Chichester, 243–274.

Whittaker A., Cope, J.C.W., Cowie, J.W., Gibbons, W., Hailwood, M.R., House, M.R., Jenkins, D.G., Rawson, P.F., Rushton, A.W.A., Smith, D.G., Thomas, A.T. and Wimbledon, W.A. 1991. A guide to stratigraphical procedure. *Journal of the Geological Society of London* **148**, 813–824.

4

Geological Time

We have seen in the previous chapter how we can interpret the relative chronology of rock units through the principle of superposition and by using cross-cutting relationships and unconformities. These tools are appropriate for limited stratigraphical studies, but in order to determine the time equivalence of units across the globe, we require an independent means of comparing rock units in relative time. Methods using faunal succession and geological events are discussed in this chapter. They have allowed geologists to construct a standard stratigraphy, known as the chronostratigraphical scale, which ensures geologists can construct an accurate picture of global Earth history. This scale does not rely upon the recognition of absolute dates (i.e. in years), but on the relative order of geological units. Absolute dates may however be determined for stratigraphical units using techniques of radiometric dating, which are discussed at the end of this chapter.

For most geologists geological time is a relative concept and involves simply placing geological units in their relative order and it is not therefore possible to talk in terms of precise ages in years. Absolute dates can be obtained but are not part of the day-to-day routine of most geological study.

4.1 Faunal and Floral Succession

Life has existed on the Earth for at least 3500 Ma, during which time it has evolved and developed irreversibly. Different species of animals (fauna) and plants (flora) have appeared, evolved and become extinct through time. The evidence for this succession of species is recorded in the fossils preserved within the rock record. This is known as the principle of faunal and floral succession and it can be used to provide relative dates for geological units, since evolution is time dependent and irreversible. We shall first consider how the early geologists developed their understanding of the fossil record and applied it in the interpretation of the geological succession.

James Hutton was able to determine crude relative chronologies and therefore interpret geological sequences within small areas, but was unable to build up a broad regional picture of the development of the geological record because he lacked the tools with which to correlate, or match, sequences of rock in one area with those in another. An independent means of correlation was needed.

This was to be provided by an English engineer William Smith (1769–1839). In 1793 he began work on the construction of a canal near Bath, in southern England. Smith built up a first hand knowledge of the strata surrounding Bath and its order of superposition. He later set out to construct the first detailed geological map of England and Wales, which was published in 1815. Smith's work was based on personal field observation and was unencumbered with unnecessary theoretical considerations. The geological units chosen by Smith for his map were based on two criteria: (1) that the lithologies or rock types were distinctive; and (2) that they contained a distinctive assemblage of fossils. The significance of Smith's observations was that for the first time, strata were identified by their distinctive fossils and that this was used as a tool in tracing map units over great distances.

In 1796, Georges Cuvier, working at the Paris Museum of Natural History, presented a paper 'On the species of living and fossil elephants' to a meeting of eminent scientists in Paris. This was a significant paper because it was the first to suggest that species of animals had become extinct in the geological past. In this paper Cuvier compared the skeletons of recently discovered fossil elephants (mammoths) from Siberia and northern Europe with the skeletons of living Indian and African elephants. He concluded that the mammoth did not belong to the same species as either of the living elephants and since there were no living representatives, it must be extinct. Cuvier was able to show with later research that this phenomenon was not isolated and that many fossils represented species that were long since extinct. Cuvier favoured catastrophic causes for the extinctions that he recognised, but it was not until the publication of Charles Darwin's *Origin of Species* in 1859 that evolution was presented as the all-encompassing theoretical basis for the successions of different fossils recognised by Smith, Cuvier and others.

Smith's discovery that strata could be identified by the fossils contained in them together with Cuvier's discovery of species extinction paved the way for the realisation that fossils could be used as indicators of relative age irrespective of sedimentary rock type. The first application of this concept was in the study of the relatively young rocks of the Cenozoic exposed in continental Europe. In the 1830s Deshayes in France, Bronn in Germany and Lyell in England all produced subdivisions of the Cenozoic that were based on fossils. In this way the foundations for the use of fossils as an independent means of correlation, and of biostratigraphy, were laid.

4.1.1   The Tools of Biostratigraphy

Biostratigraphy is the branch of stratigraphy which involves the study and interpretation of fossils to enable the correlation of sedimentary rock units.

The tools of biostratigraphy are fossils known as guide fossils (also index or zone fossils). The most useful guide fossils are those which, when living, were widely distributed both geographically and environmentally and which therefore allow a variety of rocks formed in widely spaced localities and in different environments to be correlated. As few fossil species fulfil all these ideal requirements, it follows that not all fossils are of equal value in biostratigraphy.

It is possible to draw up a set of criteria that can be used as a kind of 'check-list' to the suitability of fossils as correlation tools. Guide fossils should ideally be:

1. Independent of their environment
2. Fast evolving
3. Geographically widespread
4. Abundant
5. Readily preserved
6. Easily recognisable

Few fossils qualify for honours in all of the criteria but it follows that the more they possess the better they are as guide fossils (Figures 4.1 and 4.2).

Fossils should be independent of their environment to be of value in inter-regional correlation. Bottom-dwelling marine animals, for example, may be dependent on the nature of the sea bottom sediment (substrate) or water depth. Fossils which are strongly dependent on a narrow range of environmental parameters are of limited use as guide fossils, because they only occur in these specific environments. For example, barnacles are restricted to rocky foreshores, while corals are strongly restricted by light, salinity and temperature and both are therefore of limited value as guide fossils (Figure 4.1).

Rates of evolution are known to vary through geological time. Some fossils evolved quickly (fast-evolving) with the development of new species (species turnover) at regular intervals of less than 1 million years. High rates of species turnover, that is of evolutionary change, mean that a given fossil will exist within a smaller proportion of the stratigraphical record. For example, if a species survived for a long time then it could be found in rocks of widely different ages, but if it survived for only a short span of time its presence in a rock ties that rock to a more specific period of time. Consequently a fossil with a rapid species turnover allows the geological record to be subdivided more finely than one in which new species only appeared after long intervals. The so-called living fossils – such as *Nautilus* (the pearly nautilus), *Lingula* (an intertidal brachiopod) and *Limulus* (the king crab) – have remained unchanged in overall morphology for vast periods of time and are therefore of very limited value in biostratigraphy.

Fossils with restricted environmental niches are known as facies fossils. Despite their inadequacy some facies fossils have been pressed into service as guide fossils. Corals in particular have been used where a uniform environment is achieved over a wide area, such as in the Lower Carboniferous rocks of England. In contrast, organisms that are independent of substrate, such as those adapted to a free-swimming or floating mode of life are well-suited for correlation because they may float or swim above, and therefore drop into, a variety of different depositional environments. Such organisms are the best guide fossils (Figures 4.1 and 4.2).

Criteria / Fossil	Independent of environment	Fast to evolve	Geographically widespread	Abundant	Readily preserved	Easily recognised	Status as guide fossils
Graptolites	✓ (Plankton)	✓	✓ (Plankton)	✓	✓	✓ (Simple form)	Good (Ordovician to Silurian)
Ammonites	✓ (Free swimming)	✓	✓ (Free swimming)	✓	✓	✓ (Great diversity)	Good (Devonian to Cretaceous)
Corals	X (Need warm shallow sea)	X	X	✓	✓	✓	Poor (Carboniferous)
Echinoids	X (Bottom dwelling)	X	X	✓	✓	✓	Poor (Cretaceous)
Barnacles	X (Need rocky shore)	X	X	X	X	✓	Bad (not used)
Foraminifera	✓ (Plankton)	✓	✓ (Plankton)	✓	✓	✓	Good (Particularly Mesozoic to Recent)
Pollen	✓ (Wind blown)	✓	✓ (Wind blown)	✓	✓	✓	Good (Cretaceous to Recent)
Coccoliths	✓ (Plankton)	✓	✓ (Plankton)	✓	✓	✓	Good (Mesozoic to Recent)
Birds	✓ (Flying)	X	✓ (Flying)	X	X (Fragile bones)	✓	Bad (not used)

Figure 4.1 Examples of good and bad guide fossils. The matrix illustrates how different fossil groups match up to the ideal criteria for a good guide fossil. It is important to note that each criterion is not necessarily of equal importance. For example, preservation potential is of greater importance than widespread distribution. Bird fossils, otherwise well-suited as guide fossils, are rarely preserved and therefore, make poor guide fossils

The geographical extent of fossils obviously controls the area over which correlations can be made. Thus if an organism is world-wide in its distribution, then there is the potential for world-wide correlation of stratigraphical sequences.

The last three criteria are important because they control whether a fossil can be easily employed as a guide fossil, irrespective of whether they are naturally suited to the task. Fossils should be abundant: they will be of little use if they satisfy the first three criteria, but are impossible to find. The first bird *Archaeopteryx*, for instance, is otherwise well-suited as a Jurassic guide fossil, but only eight specimens are known to exist. Similarly, if fossils are too delicate, or their morphology too bland, or too

Figure 4.2 *Photographs of a selection of guide fossils.* **A:** *Ammonite (Lytoceras) from the Upper Jurassic of Antarctica – actual diameter 75 mm [Photograph: P. Doyle].* **B:** *Belemnite (Gonioteuthis) from the Upper Cretaceous of southern England – actual length 75 mm [Photograph: P. Doyle].* **C:** *Graptolite (Didymograptus) from the Ordovician of South Wales – actual length 35 mm [Photograph: R. Fortey].* **D:** *Foraminiferan (Neogloboquadrina) from the Holocene of the South Atlantic – actual diameter 225 μm [Photograph: F.L. Lowry].* **E:** *Trilobite (Calymene) from the Silurian of England – actual length 49 mm [Photograph: R. Fortey].* **F:** *Coccosphere (Coccolithus), composed of individual coccoliths from the present-day North Atlantic – each coccolith is 10 μm in diameter [Photograph: J. Young]*

complex, then practising stratigraphers will be unable to use them effectively. In general birds have hollow bones which are easily broken; although otherwise well-suited as guide fossils, they are limited by their poor fossil record (Figure 4.1). Graptolites preserve a simple morphology which can be recognised in even deformed rock sequences, and are excellent guide fossils (Figures 4.1 and 4.2). Fossils must therefore be readily preservable and readily recognisable to be of value.

Amongst the best fossils which can be employed in biostratigraphical correlation are ammonites: an extinct group of molluscs that died out at the end of the Mesozoic; and graptolites: an extinct group of colonial organisms common in the Palaeozoic (Figures 4.1 and 4.2). Ammonites were free-swimming organisms that had a widespread distribution and which were not tied to a particular substrate. Importantly, ammonites evolved rapidly with the production of new species and the extinction of old ones each within an estimated time span of 0.5 to 1 Ma. The practical result of this for biostratigraphy is that refined, high resolution correlation can be achieved which allows fine subdivision of the Mesozoic stratigraphical record. Graptolites were free-floating organisms which have been used for intercontinental correlation of Palaeozoic sequences (Box 4.1). Graptolites evolved quickly and, like ammonites, were not restricted to a substrate or particular type of marine environment. Both groups were abundant, readily preserved and are easily recognisable with the aid of standard reference works.

Microfossils are also particularly important in biostratigraphy, particularly as they are small enough to be extracted from the small diameter rock-cores produced from boreholes. This is obviously an advantage in the interpretation of underground sequences drilled in the quest for oil or other natural resources. Particularly important guide microfossils include foraminifera (single celled shelly organisms) and pollen (Figures 4.1 and 4.2).

Finally, it has to be said that the choice and application of guide fossils in any given stratigraphical sequence is also a function of practicality. In some intervals or locations the best guide fossils available may fall short of achieving honours in all six criteria, but may be applicable if they are abundant and widespread enough to be of value. Examples of this are the use of corals and brachiopods in the Lower Carboniferous (Mississippian), and the widespread use of echinoids in the Upper Cretaceous. Although broadly speaking these organisms are facies fossils, the

---

Box 4.1
Lapworth's graptolite biozones

In 1878 Charles Lapworth erected a series of biozones for the Lower Palaeozoic sediments (grey-wackes) of the Southern Upland hills in Scotland. These biozones were based upon graptolites, animals whose true affinities were poorly understood at that time. Lapworth used graptolites as stratigraphical ciphers which helped to unlock the story of the development of the Southern Uplands, despite the relative complexity of their geological structure, deformed by plate tectonic processes. These ciphers proved to be of immense importance and were applied widely in interpreting other sequences of similar age. Lapworth's biozones have stood the test of time and of stratigraphical fashion. Graptolites are much better understood as animals. Their planktonic lifestyle means that they are found in both deep and shallow water settings; they are exceptionally widespread in distribution, and they display rapid evolutionary development. As a result of this graptolites are ideal guide fossils. Their value is illustrated by the durability of Lapworth's original biozone scheme, which dates from the late nineteenth century, and is still used today.

Source: Fortey, R. 1993. Charles Lapworth and the biostratigraphic paradigm. *Journal of the Geological Society of London* **150**, 209–218.

uniformity of environments over a wide geographical area means that the facies has less of an impact, and the environmental dependency of the fossils is of lesser importance. In both cases, more reliable and less facies dependent fossils are uncommon and therefore broadly unworkable.

4.1.2   The Prime Unit of Biostratigraphy: the Biozone

The prime unit of biostratigraphy is the biozone, often just referred to as a zone in older literature. Biozones are strata organised into stratigraphical units on the basis of their content of guide fossils. Biozones may be recognised on local or regional scales.

The concept of the biozone was developed by Albert Oppel (1831–1865) in the 1850s. Oppel recognised that the vertical (stratigraphical) range of fossils was time significant and that it was independent of the lithology containing the fossils. He subdivided the Jurassic system into units defined on the vertical ranges of a number of fossil assemblages. In many ways this has parallels with Smith's original work, but Oppel differed in not tying the fossil assemblages to lithology. Each of Oppel's zones was formally defined and named, usually after one distinctive fossil in an assemblage, and each zone could be traced across continental Europe.

Oppel's concept of the zone, now re-christened the biozone to avoid confusion, remains valid today and some types of assemblage biozones are given the name Oppel biozones. In an assemblage biozone, the biozone is defined on the basis of the vertical ranges of a number of fossils (Figure 4.3). Often assemblage biozones are utilised where there are few good guide fossils available, as for example, is often found where most of the fossils have a benthonic (bottom dwelling) rather than planktonic (free floating) or nektonic (free swimming) mode of life. An assemblage biozone is often limited in use because it relies on the recognition of a number of fossils. Other types of biozone have also been recognised. Total range biozones are based on the total vertical range of a single fossil, usually a suitably qualified guide fossil, such as an ammonite or graptolite. Partial range biozones utilise part of the total vertical range of a fossil, particularly of organisms which were relatively slow to evolve. The partial range is usually defined as being between the first and last appearances of other fossils (Figure 4.3). Other biozones, such as the acme biozone, which is based on an abundance of a fossil group, are more difficult to recognise in practice because of the overall imperfection of the fossil record.

The lower boundaries of biozones are usually drawn at the first appearance of the guide fossil, up to the first appearance of the next (Box 4.2). These boundaries are sometimes related to an evolutionary lineage where the replacement of species is directly related to the evolution of the group from one species to another. Such biozones are called consecutive range biozones (Figure 4.3). In practice they are hard to prove and rely upon detailed stratigraphical study. In other cases the boundaries of biozones may be denoted by migrations, which mark the migration of a species from one geographical area to another and may be unrelated to the organisms of the preceding biozone. In practice, this is relatively common. The theoretical basis of any biozonation scheme is that the biological changes recorded

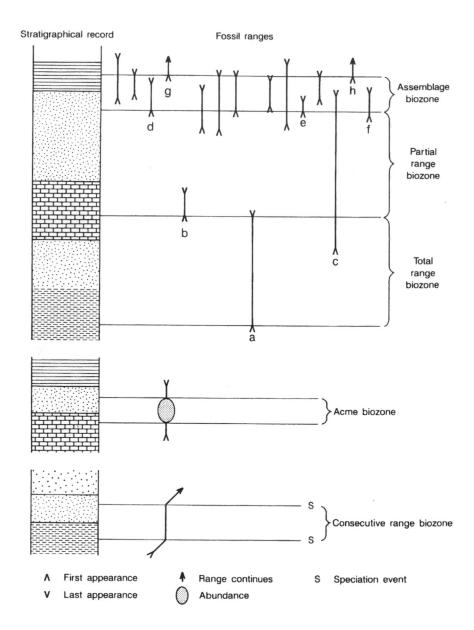

Figure 4.3   Types of biozone. The diagram shows the following types of biozone. Consecutive range biozone which is defined on the basis of the consecutive ranges of fossils in an evolving lineage. Acme biozone which is defined on the total abundance of an individual species. Total range biozone is defined by the total vertical range of species (a). The lower boundary of the partial range biozone is defined by the first appearance of species (b), and the upper boundary by the first appearance of species (d) and (e). The assemblage biozone is defined on the basis of the combined ranges of a number of fossils. [Modified from: Holland et al (1978) A Guide to Stratigraphical Procedure, Geological Society of London, Special Report No. 10, Figure 2, p. 14]

Box 4.2
Shaw's method of graphical correlation

In 1964, the American stratigrapher A.B. Shaw devised a new method of semi-quantitative correlation of biostratigraphical sequences. Shaw's method utilises a single stratigraphical section which is taken as the standard. This standard is selected because it is the thickest and most unaffected by structural complications. From this section the fossil species ranges are constructed, with particular importance being assigned to the first appearance of fossil species, but also with reference to the last appearance. A graph can be constructed using the same data collected from another stratigraphical section. The first and last appearances (bottom and top of their stratigraphical range) of the fossil taxa provide points on the graph, and a best-fit line is constructed. If a best-fit line can be constructed, the sections can be correlated (Diagram A). The technique thus provides a clear graphical method which aids in the correlation of sequences. Changes in the gradient of the graph (dog-legs) indicate changes in the rate of sedimentation (Diagram B), while gaps in the record are illustrated by plateaux (Diagram C). This method is also applicable to event stratigraphy, and both data sets can be plotted on the same curve to provide independent corroboration of correlation.

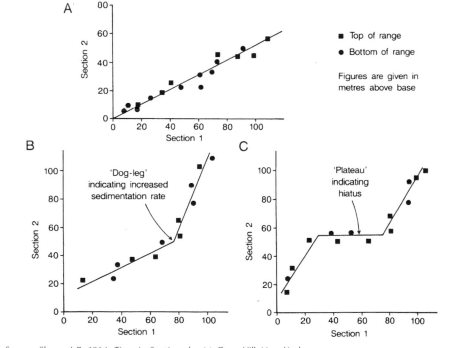

Source: Shaw, A.B. 1964. *Time in Stratigraphy*, McGraw Hill, New York.
[Diagram modified from: Shaw (1964) *Time in Stratigraphy*, McGraw Hill, Figures 20.2, 20.4 and 21.2, pp. 132, 136 and 145]

in the scheme should be synchronous across the world. A new species will not evolve and appear everywhere in an instant but will migrate progressively from a centre across a region or continent. Consequently the appearance of a new species may not be synchronous across an area. However, this is not a problem given the sheer scale of geological time and that the process of species migrations will not be detectable within the resolving power of the geological record. They can, therefore, be considered for most purposes to be instantaneous.

In terms of evolution, the concept of punctuated equilibrium, whereby species undergo periods of rapid change between long periods of stasis, has been recognised from the stratigraphical record (Box 4.3). Many zonal schemes may in fact be a reflection of such dramatic evolutionary change and may give greater relevance to the biozonal schemes based upon it, although very detailed stratigraphical study is required to prove it.

Box 4.3
Punctuated equilibrium and phyletic gradualism

In 1972, two American scientists, Niles Eldredge and Stephen Gould, wrote a paper which attempted to view evolution from the perspective of the stratigraphical record. Up until then, the accepted model of evolution was that it progressed in a gradual manner with numerous intermediates denoting evolutionary change of the entire population of a species. Eldredge and Gould called this phyletic gradualism. Darwin recognised that theoretically the morphological intermediates of an evolutionary line (a phylogeny) could be preserved in the fossil record. However, he also recognised the truth of the matter, that is, more often than not, intermediates between species (missing links) could not be found through the practical study of sedimentary successions. Darwin and many other scientists accounted for this by invoking the imperfection of the stratigraphical record, considered to be full of gaps. Eldredge and Gould questioned this on the basis of careful stratigraphical study. From their studies of Devonian trilobites and Pleistocene land snails, these authors asserted that in fact the stratigraphical record was a good one and that it supported rapid evolutionary change punctuated by long time intervals when there was no major changes in morphology. In doing this, Eldredge and Gould attacked the traditional idea of evolution by phyletic gradualism, and suggested that the stratigraphical record of unchanging morphology (stasis) followed by rapid morphological change was more illustrative of the true model of evolution, and they called this punctuated equilibrium. This model was accepted by many geologists as representative of the evolutionary changes encountered in fossil organisms, and the debate continues unabated. Recent evidence suggests that in some cases, evolution may actually represent a composite of both processes: 'punctuated gradualism'.

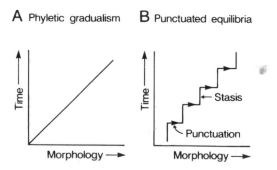

Source: Eldredge, N. and Gould, S.J. 1972. Punctuated equilibria: an alternative to phyletic gradualism. In: Schopf, T.J.M. (Ed.) *Models in Paleobiology*. Freeman, Cooper and Company, San Francisco, 82–115.

4.1.3 Lithostratigraphy and Biostratigraphy Compared: the Problem of Diachronism

As we have seen in Chapter 3 the distribution of lithostratigraphical units in time and space is determined by the nature of the depositional environment. Although the relationship of one environment to another is often linked to that of another – such as the relationship between beach, intertidal sand and offshore mud in a typical shallow marine setting – the relative positions of these environments may change through time. The deposits produced, for example during a sea level rise or fall, or during the lateral movement of a delta, will result in a lithostratigraphical unit which is internally homogeneous, but which has been formed at different times over its lateral range (Figure 3.14). In these cases, the resulting deposits transgress time boundaries and are said to be diachronous. Recognition of diachronism requires clear understanding of the faunal or floral succession which established the time lines. In this way the tie lines of correlation of diachronous units are seen to cross the time lines established by independent evidence such as guide fossils (Figure 3.12).

Diachronism is probably extremely common in the stratigraphical record, but often cannot be detected because the resolving power of biostratigraphy is insufficient. In fact, as we have seen in Chapter 3, in most cases, lithological units often develop laterally, accumulating from specific points to form a progressively continuous unit or layer. If each unit has developed laterally on a geologically insignificant scale, then it will not be detected as being time-transgressive and will normally be treated as if it was formed vertically by uniform sedimentation over its area. In such cases diachronism will be undetected by biostratigraphy. However, in large depositional environments which form over much greater periods of time, biostratigraphy provides the key to the recognition of diachronism.

One of the best examples of a large-scale diachronous deposit is that of the Lower to Middle Jurassic yellow sandstones of southern England, which crop out continuously from Dorset to the Cotswolds. These sandstones have a distinct colour and mineral content, forming an extremely distinctive and internally homogeneous deposit which belongs to one lithostratigraphical unit (a formation). However, detailed study of the ammonite faunas of these sandstones at a variety of localities shows the sand body to be diachronous (Figure 4.4). In the Cotswolds the sandstones contain ammonites indicative of a much older biozone than those present in the same sandstone units of identical lithology in Dorset. Sedimentological models suggest that these sandstones formed on a shallow shelf as a sand-bar or sheet which migrated progressively southwards in response to tidal or current activity. Although the detail of this model has been questioned, it illustrates the time-transgressive nature of a deposit which has a homogeneous lithological character (Figure 4.4).

In summary, biostratigraphy provides the key to the relative age of lithostratigraphical units. Biostratigraphical units have boundaries that are time significant and isochronous (of equal age), while lithostratigraphical units have boundaries based only on lithology and may therefore be diachronous.

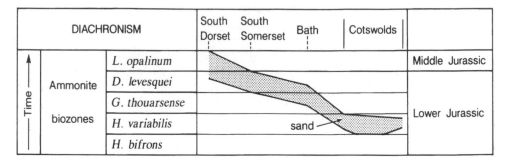

The sand horizon was laid down at different times in
different regions and is said to be DIACHRONOUS

Figure 4.4   *Diachronism: an example from the Lower Jurassic sandstones of southern England. The sandstone unit gets progressively younger moving from the Cotswold Hills to Dorset, as shown by the ammonite biozones. [Modified from: Rayner (1981)* The Stratigraphy of the British Isles, *Cambridge University Press, Figure 61, p. 81]*

4.2   Events in the Geological Record

The sedimentary record is often thought to have been formed by gradual accumulation over a significant period of time. Closer examination of present-day processes and environments shows, however, that some deposits are actually deposited relatively quickly and may be considered as instantaneous when measured against the vast scale of geological time. These deposits can often be related to specific depositional events, such as a volcanic eruption or a severe storm. By their nature these event horizons are time significant and important as an independent means of providing relative dates for the stratigraphical succession.

James Hutton's principle of the uniformity of process, now termed actualism, allows geologists to deduce the nature of the agencies involved in the production of a lithology from direct observation of similar deposits which are forming today. In the 1930s it was discovered that submarine canyons were carved in the slopes of continental shelves as a result of catastrophic flows of sediment. For example, on 18 November 1929, at Grand Banks, Newfoundland, one such flow sequentially cut transatlantic telephone cables over a period of 13 hours, which allowed the velocity of the flow to be calculated at over 45 knots. This is a dramatic event, which geologically speaking was instantaneous and which deposited a large body of sediment known as a turbidite. This deposit can be said to have been both rapidly deposited, and associated with a specific event (the turbidity current). As another example, it has been shown recently that severe storms or giant waves (tsunamis) have had a significant impact on the sedimentological record and produce deposits which are effectively created instantaneously (Box 2.1). Many other examples of catastrophic events have been recorded in the geological record and provide important time horizons.

4.2.1   Event Stratigraphy

The products of catastrophic events are geologically instantaneous and therefore effectively represent punctuations of the stratigraphical column which can be used to denote time planes. Such isochronous horizons are known as event horizons and their study is that of event stratigraphy.

The value of event horizons lies in the fact that they are deemed to have been produced at the same instant over the area that they were deposited. Some deposits may be genuinely instantaneous, with timescales of hours or minutes, or alternatively they may have accumulated over some time following a short-lived event, such as in the fall-out of volcanic dust after an eruption (Figure 4.5). In an eruption, the coarser particles are deposited first, while the finer fractions take time to be deposited out of suspension. However, in geological terms such time- averaging of events is negligible and difficult to detect. Geologically therefore such horizons (e.g. ash horizons) can be considered to be isochronous (i.e. the same age everywhere) and, if present over a wide area, can be used to correlate one geological section with another. In some cases it has been shown that correlation can be achieved to a greater resolution than that of the best biozones. Some event horizons, such as that produced by the fall-out of volcanic ash, have the advantage of being easily recognisable in the rock record and may cover a large area. Ash may be deposited in a wide variety of different environments (e.g. at sea, on land, or in lakes) providing the potential to correlate between very different depositional environments, something which is difficult to do with fossils, because there are few fossils common to all environments (e.g. marine animals can not live on land). Tephrostratigraphy is the study of such volcanic

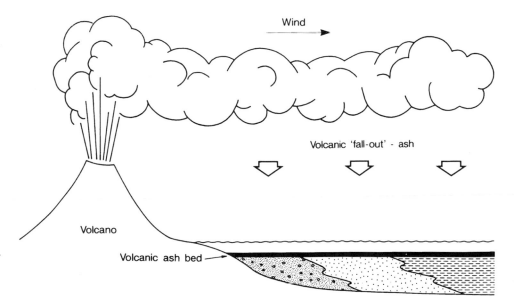

Figure 4.5   *Tephrostratigraphy: an example of event stratigraphy. The layer of volcanic ash is deposited across several different depositional environments or facies and provides an isochronous horizon*

deposits and is widely used as a correlative tool, since each ash horizon usually has a distinct geochemical signature.

Events can be classified by their origin into either: (1) physical events, (2) chemical events, (3) biological events and (4) composite events.

Obvious examples of physical event horizons are tephra or ash layers produced by volcanic eruption. The eruption of Mount St Helens in 1981 produced large volumes of tephra which will eventually be incorporated into the stratigraphical record as a physical event horizon. The products of storms, tsunamis, meteorite impacts and mass movements all provide examples of physical event horizons. Less dramatic but more regular physical events are produced by the reversal of the Earth's magnetic field probably caused by switching of the electrical and fluid current flows in the Earth's core. These reversals impose a magnetic fingerprint on sedimentary and igneous rocks, because any magnetic particles within them align themselves with the prevailing magnetic field and become locked into the rock as it is lithified or as it cools. These magnetic events are a powerful tool in correlation in their own right and their study is often referred to as magnetostratigraphy. Figure 4.6 illustrates the application of magnetostratigraphy in the correlation of seven deep sea cores from the Southern Ocean. Each of these cores contains very different sediments, ranging from diatomaceous oozes to calcareous silts. They can, however, be correlated precisely on

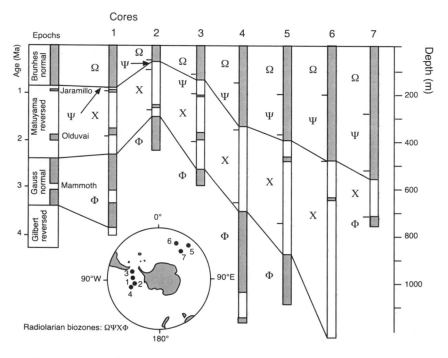

Figure 4.6 *The application of magnetostratigraphy. Seven marine cores from the Southern Ocean of varying lithology are correlated on the basis of their palaeomagnetism. Note the close correlation between the Radiolarian biozones and the palaeomagnetic time lines. [Modified from: Opdyke et al (1966)* Science **154**, *Figure 1, p. 350]*

the magnetic properties of these different sediments which contain a record of the Earth's recent magnetic reversals. Many geologists consider magnetostratigraphy to be a branch of stratigraphy in its own right, but it is effectively related to specific events in Earth history and is therefore a part of event stratigraphy.

Chemical event horizons are often difficult to identify without detailed technical equipment, but usually take the form of 'spikes' in the geochemical signature of lithological units which illustrate unusual concentrations of particular elements or stable isotopes (Figure 4.7). Chemostratigraphy is a term often used to refer to the use of such event horizons in stratigraphy. Box 4.4 illustrates one example of chemostratigraphy, in this case the use of strontium isotopes.

Biological events are mostly associated with the rapid colonisation of particular environments, usually on a small geographical scale, or are otherwise associated with extinction events. Often, events may be composite in origin, combining physical, chemical and biological characteristics. For example, the eruption of Mount St Helens is obviously a composite event: the tephra has a physical presence as well as a geochemical signature and the destruction of the surrounding forest was a biological event.

Perhaps the most topical events in the geological record are mass extinctions. These events are effectively dramatic reductions in the diversity of animal and plant life.

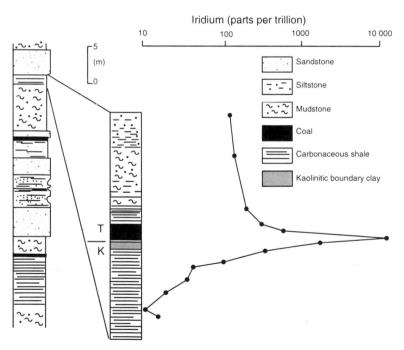

Figure 4.7   *A chemical event horizon. The element iridium is concentrated into a distinct layer, or 'spike', at the Cretaceous–Tertiary boundary in this section from New Mexico. The iridium spike is probably a product of a meteorite impact (see Box 10.4). [Modified from: Kauffman (1988)* Annual Review of Earth & Planetary Sciences **16**, *Figure 7, p. 624]*

Box 4.4
Strontium Isotopes: an example of chemostratigraphy

In recent years the use of stable isotopes has introduced a range of new tools into the strati-graphical tool kit. Stable isotopes are elements who my have a different number of neutrons in their nucleus thereby giving them a different atomic mass, but unlike unstable isotopes are not subject to radioactive decay. Strontium isotopes are stable and valuable in chemostratigraphy. The method relies on the measurement of an $^{87}Sr/^{86}Sr$ value in a sample taken from biogenic calcite in a marine fossil; this ratio expresses simply the relative numbers of atoms of the isotopes of strontium that have mass numbers of 86 and 87. On precipitation of biogenic calcite some strontium is incorporated from seawater. The strontium in the calcite will have an $^{87}Sr/^{86}Sr$ ratio equal to that of the sea water at that time it was precipitated. As the $^{87}Sr/^{86}Sr$ ratio has changed in the world's oceans in a known way, comparing the $^{87}Sr/^{86}Sr$ value of a fossil with a standard curve permits an age for the fossil to be deduced, or the fossil can be correlated with a fossil in a different section, but with the same $^{87}Sr/^{86}Sr$ value. The strontium isotope ratio in sea water is controlled by the circulation of water through mid-ocean ridges, by input from rivers, and by the recrystallisation of carbonate sediments during burial; since the geometry of the ocean basins, continents and mid-ocean ridges have changed with time so has the ratio. By determining the ratio in fossils of known age, dated by independent methods, a standard curve can be obtained. This technique is illus-trative of the increasing technical sophistication of chemostratigraphy and its increasing role within stratigraphy as a whole.

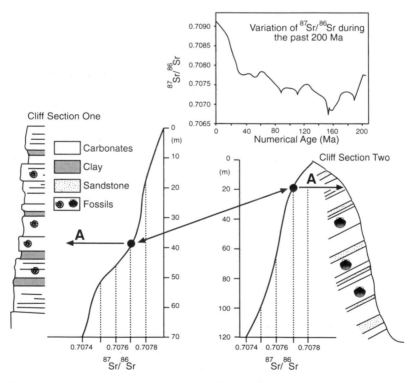

Source: McArthur, J.M. 1998. Strontium isotope stratigraphy. In Doyle, P. and Bennett, M.R. (Eds) *Unlocking the Stratigraphical Record.* John Wiley and Sons, Chichester, 221–241. [Diagrams modified from: McArthur (1998) *Unlocking the Stratigraphical Record.* Wiley, Figures 8.1 and 8.3, pp. 222 and 228.]

Mass extinctions are examples of biological events which may be caused by physical events, such as meteorite impacts. They occur over an extremely short space of geological time and many may be attributed to the catastrophic effects of changes in climate and sea level which destabilised the Earth's ecological balance. The most famous example of a mass extinction is that which occurred at the end of the Cretaceous period. In fact this event horizon marks the boundary between the Mesozoic and Cenozoic erathems. Recent work has attempted to explain this event in terms of a global catastrophe caused by the impact of an extra-terrestrial object, probably an asteroid, of over 10 km in diameter. The evidence for this lies in the recognition of a thin layer of clay at many sites across the globe, in which there is an abnormal enrichment in the element iridium (Figure 4.7). On Earth, iridium is concentrated in the core and is uncommon in the crust. Iridium is, however, common in meteorites and other extra-terrestial bodies. This layer of clay acts both as a chemical and physical event horizon. Other explanations for this mass extinction have been proposed and are discussed in Section 10.4.1.

In summary, catastrophic events within the geological record provide important horizons or layers with which to correlate rock sequences between different areas which experience the same event. The power of the tool lies in the fact that the geological record of most catastrophic events is effectively isochronous.

4.3   The Chronostratigraphical Scale

One of the most important goals in stratigraphy is establishing the global standard or chronology of geological units. It should be possible to correlate the rocks of given areas with this global scale, so that geologists wherever they may be working, can locate the rocks they are studying within Earth history. This global scale is known as the Chronostratigraphical Scale and has been established through the work of numerous geologists over the last 150 years or so. The Chronostratigraphical Scale is a summation of all stratigraphical knowledge and as such there is no single section on the surface of the Earth at which all its units are exposed. The Chronostratigraphical Scale consists of chronostratigraphical units.

Chronostratigraphical units are bodies of strata that were formed during specific periods of geological time. The boundaries of chronostratigraphical units are time-significant as they are of the same age across the globe (i.e. they are isochronous), and dates in years can be applied to these boundaries through the application of the radiometric dating techniques discussed in Section 4.4 below. The Chronostratigraphical Scale provides the global standard with which local sequences may be correlated. A simplified version of the currently accepted scale is given in Figure 4.8. This standard allows a uniformity of approach to be achieved by geologists world wide and is agreed through international co-operation.

Chronostratigraphical units are sometimes referred to as time-stratigraphical units to distinguish them from rock-stratigraphical units (lithostratigraphical units). Chronostratigraphical units are in essence the repository for all stratigraphical knowledge. The mechanisms of the chronostratigraphical procedure are discussed in Box 4.5. The Chronostratigraphical Scale is a standard with which geologists

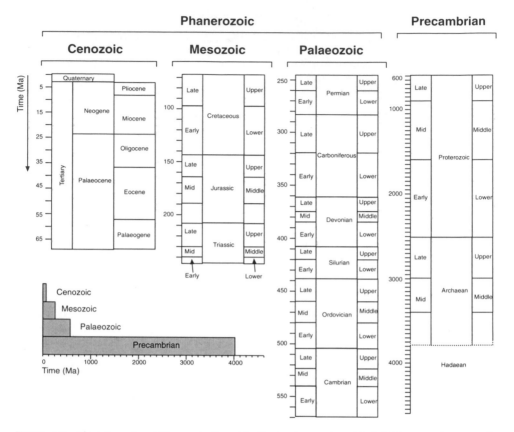

Figure 4.8   *The International Chronostratigraphical Scale, chrostratigraphical divisions are shown on the right of each column, the equivalent time units are shown on the left. [Based on information in: Palmer (1983) Geology **11**, p. 508]*

can correlate their individual rock sequences and consists of a series of systems (e.g. the Carboniferous System or Triassic System). Each system consists of rocks deposited during the same time interval. The main systems were identified for the most part over a period of 50 years in the first part of the nineteenth century. As illustrated in Figure 4.9 most of the systems were established from the study of the stratigraphical record in Europe and were initially defined on lithology alone. The development of the Chronostratigraphical Scale as we know it today can perhaps be attributed to Murchison, who in 1835 defined his Silurian System on the basis of its fossil content and thereby paved the way for the recognition of major rock bodies with time significant boundaries. The development of the Chronostratigraphical Scale is seen by some scientists to be one of the most significant achievements in geology and allows an independent and time-significant yardstick with which to measure the geological record.

Today most of the boundaries of chronostratigraphical units are determined using biostratigraphy, although other techniques are increasingly being used. Since there is no locality at which all the chronostratigraphical units which make up the global scale

Box 4.5
Mechanisms of chronostratigraphy

The basis of the concept of chronostratigraphy is that it acts as a repository for stratigraphical information. Chronostratigraphical units are bodies of rock whose boundaries are time-significant, although not necessarily lithologically distinct from the units above and below. As a consequence, international agreement is needed to determine where the boundaries of global reference sections should lie and at that point a symbolic 'golden spike' is driven in. Correlation of other rock sequences with the global standard is undertaken on the basis of independent relative dating methods such as biostratigraphy or event stratigraphy, but not usually lithostratigraphy. Absolute dates may be obtained for the duration of chronostratigraphical units through the application of radiometric dating.

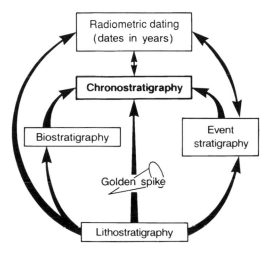

Source: Holland, C.H. 1986. Does the golden spike still glitter? *Journal of the Geological Society of London* **143**, 3–21.

are exposed, individual boundaries are identified through stratotype sections located across the world. In other words the best exposed example of the boundary between each unit is selected as typical, and designated as a stratotype section, usually through international agreement This allows geologists to correlate their individual sections with a known, protected, locality which they can visit. The systems were originally defined largely in geological and geographical terms and as a consequence the boundaries were not always fixed, often causing confusion. More recently the boundaries have been decided by international agreement and a symbolic spike, known as a golden spike, is driven in to the chosen stratotype section, to mark the agreed position of the isochronous boundary.

The systems may be grouped into larger units, known as erathems. These too are defined by faunal changes. John Phillips (1800–1874), the nephew of William Smith, subdivided the part of the geological record which contained fossils into three main subdivisions based on major changes in the fauna (Figure 4.10A). These erathems

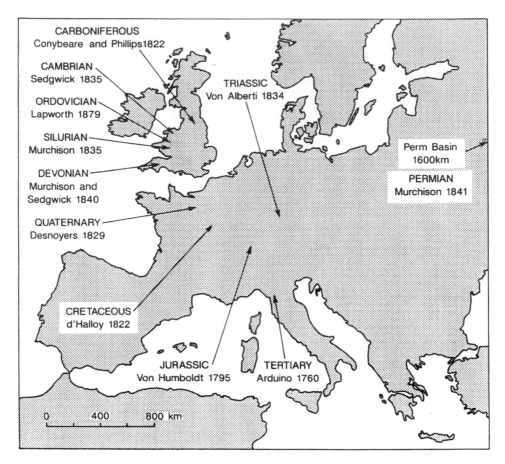

Figure 4.9 *Geographical regions in which key geological 'Systems' were first defined. [Modified from: Eicher (1976) Geologic Time, Prentice-Hall, Figure 3.2, p. 59]*

have their boundaries marked by mass extinction events. Phillips named them with respect to the nature of their fauna: Palaeozoic (old life), Mesozoic (middle life) and Cainozoic (new life: now commonly called the Cenozoic), reflecting increasingly familiar fossil organisms through time to the present. The Palaeozoic–Mesozoic boundary is denoted by the Permian–Triassic mass extinction and the Mesozoic–Cenozoic by the Cretaceous–Tertiary mass extinction. The boundary between the older primary or Precambrian rocks and the Palaeozoic was taken at the first appearance of shelly fossils, also now interpreted as a biological event horizon associated with the diversification of life on the planet. It is worth emphasising that the major subdivisions of the Chronostratigraphical Scale are bounded by biological event horizons, usually extinctions or radiations (Figure 4.10B).

Recently it has become the practice to distinguish between chronostratigraphical units and geological time (Figures 4.8 and 4.11). Chronostratigraphical units are those bodies of rock deposited during a specific period of geological time. The actual time elapsed may only be measured by the application of radiometric dating which is

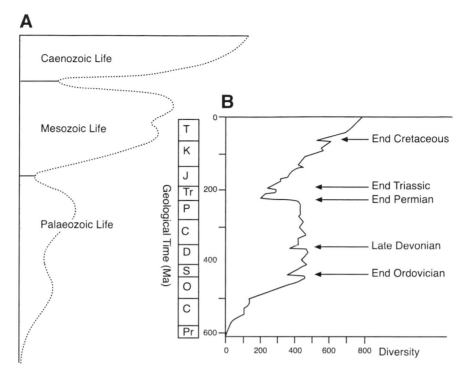

Figure 4.10 *Faunal diversity through geological time.* **A:** *Phillips' Erathems, based on faunal diversity. The area to the left of the dashed line represents the diversity of life; the boundaries of the erathems are drawn at intervals of major falls in diversity of life.* **B:** *The five major periods of extinction during the Phanerozoic. During these intervals, diversity has declined rapidly. The most dramatic event was in the Permian. The diagram also illustrates that there were many minor extinction events. [Modified from:* **A:** *Phillips (1860) Life on the Earth: its Origin and Succession, Figure 4, p. 56.* **B:** *Erwin et al (1987) Evolution* **41**, *Figure 1, p. 1178]*

discussed below. Geological time is an abstract concept and there is no guarantee that the chronostratigraphical record is truly representative of the whole of the amount of geological time elapsed. There may, for example, be periods in Earth history when no rocks were deposited anywhere. Geologists make the distinction because it is useful to have time terms (e.g. Early Carboniferous) to be able to refer to historical events and circumstances and chronostratigraphical terms (e.g. Lower Carboniferous) to refer to the relative position of stratigraphical units. The equivalent chronostratigraphical and geological time terms are given in Table 4.1.

4.4   Absolute Geological Time

Absolute dating is distinguished from relative dating as it deals with absolute numbers rather than the simple ordering of sequences based on fossils and other data. It relies for the most part upon the application of radiometric techniques.

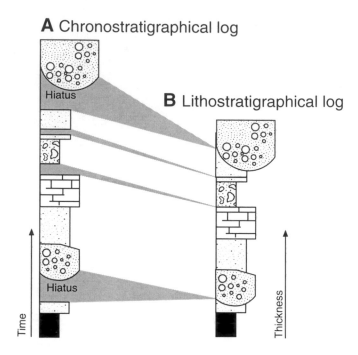

A Chronostratigraphical log

B Lithostratigraphical log

Hiatus

Hiatus

Time

Thickness

Figure 4.11   *The relationship of chronostratigraphy and lithostratigraphy. The same rock succession is plotted in (A) with time as the vertical axis, and in (B) with thickness as the vertical axis. Thickness is not a true indicator of time elapsed, because there are periods of non-deposition (hiatuses) which are clearly illustrated on the chronostratigraphical log.*

Table 4.1   *Chronostratigraphical units and their equivalent geological time units*

Chronostratigraphical terms	Geological time terms
Eonothem (e.g Phanerozoic Eonothem)	Eon (e.g. Phanerozoic Eon)
Erathem (e.g. Palaeozoic Erathem)	Era (e.g. Palaeozoic Era)
System (e.g. Carboniferous System)	Period (e.g. Carboniferous Period)
Series (e.g. Lower Carboniferous Series)	Epoch (e.g. Early Carboniferous Epoch)
Stage (e.g. Visean Stage)	Age (e.g. Visean Age)

The possibility of using naturally occurring radioactive elements to date rocks was first suggested by the physicist Lord Rutherford in the early twentieth century. Arthur Holmes was the first geologist to construct a geological time scale based on radiometric dating and this has formed the basis of the refined high resolution scale available today (Figure 4.8). The method applied then, as now, is based upon the radioactive decay properties of unstable isotopes. These unstable isotopes break down to more stable ones by emitting atomic particles and/or energy. This radioactive decay is time-dependent and forms the basis of radiometric dating.

4.4.1   The Principles of Radiometric Dating

The nucleus of an atom contains positively-charged mass particles called protons and mass particles with no electrical charge known as neutrons. The nucleus of an atom occupies only about 10^{-4} of the volume of the atom but contains essentially all the mass. Orbiting the nucleus of an atom at various distances are particles known as electrons, the outermost of which control the chemical properties. Chemical elements are classified according to the composition of the nucleus. Two terms are used: (1) the atomic number, which is the number of protons within a nucleus; and (2) the atomic mass number which is the number of protons plus neutrons. It is the convention to give the numerical value of the atomic number as a superscript and the atomic mass number as a subscript on the left hand side of the symbol for a chemical element. For example, the element ^{238}Uranium has a nucleus with 92 protons, 146 neutrons and therefore has total of 238 particles in its nucleus:

$$^{238}_{92}U$$

For an individual element the atomic mass number can vary, since the number of neutrons in the nucleus is not always constant. Elements having the same number of protons, but a different number of neutrons are called isotopes. For example, oxygen has two isotopes. Both have the same number of protons (8), but one has 8 neutrons and the other 10, which gives ^{16}O and ^{18}O. Each isotope of an element is called a nuclide. Certain, although not all, isotopes are unstable and will alter through time by the emission or sometimes capture of atomic particles together with electromagnetic radiation to obtain a more stable form. This alteration is known as radioactive decay and the isotopes involved are known as radioactive nuclides. In radioactive decay, the original isotope is termed the parent nuclide and the product of the decay is called the daughter nuclide. The value of this process to geology is that the rate of radioactive decay is characteristically exponential and time dependent. The time scale of decay is often expressed by the half life of the nuclide. This is defined as the period of time required to reduce a given number of parent nuclides by one half, and this is constant for each isotope. For example, if 100 atoms of a parent nuclide are left to decay, after one half life only 50 of the parent atoms will remain. It will then take the same period of time to reduce the remaining 50 parent atoms to 25 and so on.

Given these principles, an absolute radiometric date can be obtained from a specific radioactive nuclide so long as the following information is obtained: (1) the proportion of the parent nuclide present; (2) the proportion of the daughter nuclide present; and (3) the rate at which decay occurs or the half life. Given these three pieces of information an absolute age in years for a rock can be obtained. Ideally for highly accurate age dating certain conditions should be met: when a radioactive nuclide first forms or becomes incorporated into a rock none of its non-radiogenic nuclides must be present; no parent or daughter nuclides must be added or removed from the rock being dated, in other words the system must be a closed one. In practice these ideal conditions rarely occur but deviation from them can be readily corrected for, enabling the calculation of quite accurate absolute age estimates.

4.4.2   Methods and Limitations of Radioactive Dating

A variety of different radioactive nuclides and methods can be used for dating geological events as shown in Figure 4.12. The magnitude of the half life helps determine the length of time over which each method is appropriate: the shorter the half life the shorter the time period over which reliable data can be obtained.

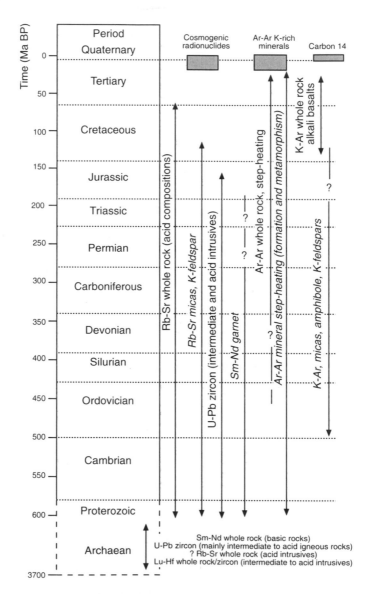

Figure 4.12   *Different methods of radiometric dating and their applicability to different rock types and time periods. [Reproduced with permission from: Hole (1998) In: Doyle and Bennett (Eds) Unlocking the Stratigraphical Record, John Wiley & Sons, Figure 12.11, p. 380]*

For example, carbon 14 dating can only usefully be used to provide dates during the later part of the Quaternary as its half life is only 5730 years; in contrast Uranium 238 has a half life of 4469 Ma and can therefore be used to date Precambrian rocks.

Methods of dating may either be based on individual minerals or on whole rock samples. Individual minerals provide information about the age at which mineral formation and crystallisation occurred. This may reflect the original formation of the mineral as it cooled in an igneous rock body or its subsequent recrystallisation during metamorphism (metamorphic resetting). In rocks which have been subject to only low grade metamorphism it is possible to distinguish between these two ages by undertaking both mineral and whole-rock measurements. Information about the

---

Box 4.6
Radiometric age bracketing: an example from the British Quaternary

Radiometric dates within the recent Quaternary record are mainly based on Carbon 14 dating. Evidence of a glacial advance is recorded by the presence of a glacial till; unfortunately a glacial till cannot be dated directly. It can, however, be dated through age bracketing. As a glacier or ice sheet advances to deposit the till, organic remains may be overridden and collect in pockets beneath the till layer. Similarly, as the glacier decays, organic material may collect in hollows on the surface of the till. Consequently, at exceptional locations one may find a till which is bracketed top and bottom by organic remains which can be dated by radiocarbon assay. The dates from the organic material above and below the till provide an upper and lower age for the till layer and therefore provide an estimate of the duration of the glacial advance. The best age control is provided by dating the top-most layer of the basal organic material and the bottom-most layer of the organic material above the till.

Age bracketing has been used at Dimlington in Holderness to provide age limits for the main glacial advance of the Devensian glaciation in that area. The Devensian glaciation was the last main ice advance of the Quaternary 'Ice Age' to cover much of the British Isles. The age constraint provided at Dimlington is sufficiently good for this site to be regarded by most people as the Quaternary type section for this glacial advance.

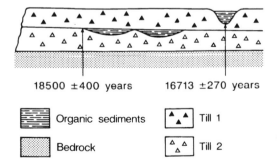

18500 ±400 years   16713 ±270 years

Organic sediments   Till 1

Bedrock   Till 2

Till 1 was deposited sometime between
18500 ±400 years and 16713 ±270 years
based on the two radiocarbon dates

Source: Rose, J. 1985. The Dimlington Stadial/Dimlington Chronozone: a proposal for naming the main glacial episode of the Late Devensian in Britain. *Boreas* **14**, 225–230.

thermal history of a sample can also sometimes be obtained using Argon/Argon dating which involves step-by-step heating of the mineral or rock sample. Dates obtained at the highest temperatures reflect the original mineral formation, while those obtained at lower temperatures may reflect subsequent thermal events in the life of the mineral or rock. When dealing with very old igneous rock bodies from the Early Palaeozoic which have been subject to very complex thermal histories, associated with, for example, multiple episodes of orogenesis, metamorphism, compaction and burial, these issues are of particular importance.

An obvious limitation of radiometric dating is that in most cases it is restricted to dating crystalline rocks (igneous or metamorphic). This is because both rock groups represent relatively closed systems, in which the radioactive clock is set at the time of crystallisation of their constituent minerals. In contrast, with very few exceptions, any radiometric date derived from sedimentary rocks will give only the ages of the source rocks from which their minerals were derived by erosion of crystalline rocks, unless the mineral itself formed at the time of formation of the sediment, a relatively rare case. However, carbon 14 dating of organic matter in sediments is one exception to this rule. The unstable carbon 14 nuclide is constantly being formed in the upper atmosphere through the bombardment of the stable nitrogen 14 nucleus by cosmic rays. This radioactive carbon rapidly cycles through the atmosphere into the hydrosphere and the biosphere. Once locked into sediments in the form of organic material it behaves essentially as a closed system, its clock starting from the time of incorporation in the sediment. With a half life of only 5730 years, the nuclide can only be used to date quite young rocks (i.e. from the Late Quaternary).

Sedimentary sequences may be dated, however, through the use of age bracketing. For example, in a sequence of sedimentary rocks which are interbedded with lava flows it is possible to date the lava flows and use them to provide age brackets for the sedimentary rocks between them. In younger rocks this technique can be used where there is organic material in the layers above and below that to be dated, using carbon 14 dating. This technique of carbon 14 age bracketing is common when dating Quaternary sediments and is illustrated in Box 4.6.

4.5   Summary of Key Points

- Biostratigraphy is the application of fossils to stratigraphy, in particular as an independent means of correlation of lithostratigraphical sequences.
- Fossils useful in biostratigraphy are known as guide or index fossils. Not all fossils are appropriate as guide fossils. Guide fossils should be independent of environment, widespread, fast-evolving, abundant, readily preserved and readily recognisable.
- The fundamental unit of biostratigraphy is the biozone, a body of strata distinguished by its fossil content.
- Biostratigraphical units are fundamentally time-significant, independent of lithology. Lithostratigraphical units are not inherently time-significant and are dependent on lithology.

- Event stratigraphy is the recognition of the products of geologically instantaneous events in the geological record and the use of them in correlating sequences. Event horizons are time-significant and isochronous.
- Events may be physical (e.g. volcanic eruptions), chemical (e.g. regional change in water chemistry), biological (e.g. extinctions) or composite.
- Chronostratigraphical units are bodies of strata that are bounded by isochronous surfaces and which are therefore time-significant. Chronostratigraphical units include the systems (e.g. Carboniferous System), defined mostly on biostratigraphical grounds. Event horizons provide an additional means of defining chronostratigraphical unit boundaries.
- The Chronostratigraphical Scale of chronostratigraphical units exists as a means of international correlation.
- The dating of geological events in years is known as absolute dating. Absolute dating relies on the known radioactive decay of certain unstable isotopes.

4.6 Suggested Reading

Eicher's book *Geologic Time* (1976), although quite old provides a very nicely written and readable account of the principles of time and correlation in stratigraphy: Chapter 6 contains a clear discussion of the basis of radiometric dating. Snelling (1987) provides a useful discussion of the chronological subdivision of the geological record. Whittaker et al (1991) provides clear definitions of the biozones and of biostratigraphy and their relationship to event stratigraphy and chronostratigraphy, and this is also dealt with in depth by Hedberg (1976). Kauffman and Hazel (1977) is a collection of advanced papers on biostratigraphy while Kauffman (1988) provides an in-depth review of the concepts behind event stratigraphy. Hailwood and Kidd (1993) is a collection of rather advanced research papers on the use of various methods of determining relative time in stratigraphical sequences. Hole (1998) provides an advanced review of some of the principles and techniques of radiometric dating, while practical examples of the selection and use of guide fossils are provided for the Mesozoic by Kennedy and Cobban (1976) and Doyle and Bennett (1995).

References

Doyle, P. and Bennett, M.R. 1995. Belemnites in biostratigraphy. *Palaeontology* **38**, 815–829.

Eicher, D.L. 1976. *Geologic Time*, (Second Edition). Prentice-Hall, Engelwood Cliffs, New Jersey.

Hailwood, E.A. and Kidd, R.B. (Eds) 1993. *High Resolution Stratigraphy*. Geological Society Special Publication No. 70, Geological Society, London.

Hedberg, H.D. (Ed.) 1976. *International Stratigraphic Guide: a Guide to Stratigraphical Classification, Terminology and Procedure*. John Wiley, New York.

Hole, M.J. 1998. Stratigraphical applications of radiogenic isotope geochemistry. In Doyle, P. and Bennett, M.R. (Eds) *Unlocking the Stratigraphical Record*. John Wiley, Chichester, 351–382.

Kauffman, E.G. and Hazel, J.E. (Eds) 1977. *Concepts and Methods of Biostratigraphy*. Dowden, Hutchinson and Ross, Stroudsberg.

Kauffman, E.G. 1988. Concepts and methods of high-resolution event stratigraphy. *Annual Reviews of Earth and Planetary Science* **16**, 605–654.

Kennedy, W.J. and Cobban, W.A. 1976. Aspects of ammonite biology, biogeography and biostratigraphy. *Special Papers in Palaeontology* **17**, 1–94.

Snelling, N.J. (Ed.) 1987. *The Chronology of the Geological Record.* Geological Society Memoir No. 10, Geological Society, London.

Whittaker A., Cope, J.C.W., Cowie, J.W., Gibbons, W., Hailwood, M.R., House, M.R., Jenkins, D.G., Rawson, P.F., Rushton, A.W.A., Smith, D.G., Thomas, A.T. and Wimbledon, W.A. 1991. A guide to stratigraphical procedure. *Journal of the Geological Society of London* **148**, 813–824.

5

Interpreting the Stratigraphical Record

So far we have examined the tools that allow us to determine the stratigraphical succession and age of rock units. In this chapter we will now examine the tools that allow us to interpret stratigraphical units as the product of the environment in which they formed. From this we can make deductions about the nature of the Earth's environment through geological time.

5.1 Facies and their Interpretation

Sedimentary facies are bodies of sedimentary rock that are the product of a particular depositional environment. We can often determine this environment by examining a set of characteristics, which taken together are indicative of that environment. The term facies is not a concept which is exclusive to sedimentary environments, but can also be applied in igneous and metamorphic terrains. Here the sum total of the characteristics of these rock bodies allows geologists to determine the conditions existing during their formation (Box 5.1).

The concept of sedimentary facies was first described by Armanz Gressly (1814–1865) in 1838 while working on the Jurassic rocks of Switzerland. Gressly was concerned for the most part with the relationship between changes in fossil content and variation in lithology. He was convinced that certain fossils were linked to certain rock types and that together fossils and lithology could help determine the nature of the environment in which the rocks were deposited. Many other geologists were

Box 5.1
Barrow zones

The application of the facies concept to mapping metamorphic rocks was explored by Barrow (1893) in a pioneering stratigraphical study of the south-east Grampians of Scotland. Within this area fine grained meta-sedimentary (pelitic) rocks of the Dalradian sequence crop out. Barrow mapped a series of boundaries within these rocks on the basis of the first appearance of certain characteristic metamorphic minerals. He identified five metamorphic 'zones' of pelitic rock by this technique: chlorite, biotite, garnet, kyanite and sillimanite. The extent of each of these zones is shown in the diagram below. In doing this, Barrow was simply applying stratigraphical techniques to mapping metamorphic rocks which showed changing mineral assemblages. Later experimental work showed that the sequence of zones represented progressively increasing pressure and temperature conditions, controlled by depth of burial during regional metamorphism (Chinner 1966). The work of Barrow and others therefore provides an insight into the environmental conditions of metamorphism and consequently the tectonic history of the region.

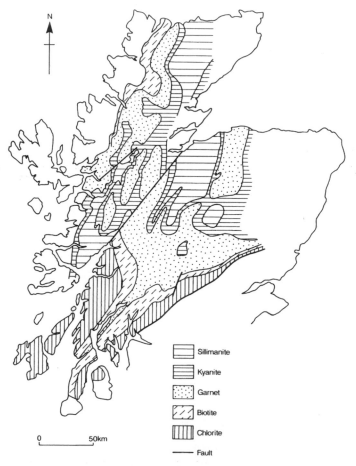

Sources: Barrow, G. 1893. On an intrusion of muscovite-biotite gneiss in the south-eastern Highlands of Scotland and its accompanying metamorphism. *Quarterly Journal of the Geological Society of London* **49**, 330–335. Chinner, G.A. 1966. The distribution of pressure and temperature during Dalradian metamorphism. *Journal of the Geological Society of London* **122**, 159–186. [Diagram modified from: Winchester (1974) *Journal of the Geological Society of London* **130**, Figure 3, p. 514.]

making similar observations and inferences at this time, but it was Gressly who first coined the term facies to encompass the sum total of all the characteristics of a rock body which allow one to determine its origin.

Actualism is the theoretical basis for facies analysis and it entails the detailed study of modern environments, to determine not only the processes by which deposition occurs but also the products of those processes. In the present day it is possible to observe processes which result in a product and the causal link between the two can be established. In the stratigraphical record we have only the product. Analysis of the characteristics of that product and comparison with the products of modern depositional environments is the key to their correct interpretation. This process is the basis for most of the deductive reasoning behind the study of facies (Figure 5.1).

The characteristics of individual sedimentary facies can be determined under four headings: (1) geometry, (2) lithology, (3) sedimentary structures and (4) fossils.

The geometry of a sedimentary rock body is a function of the environment in which it was deposited. River channels, sand dunes and beaches all have a characteristic shape and the sedimentary rock bodies formed in these environments will therefore have a similar geometry. Observing geometry is therefore an important first step in understanding the processes governing the formation of a particular facies.

Lithology is extremely important and careful observation of all lithological characteristics is necessary in order to deduce the type of environment in which it was deposited. Of particular importance are colour, texture (i.e. grain size, grain shape and grain roundness) and composition. In desert environments for example, iron within sediments is exposed to the atmosphere and is oxidised, which creates iron oxides (i.e. it 'rusts'). Oxidised iron minerals are known as haematite, a mineral characterised by its dark red colour. In this way red sedimentary bodies, known as red beds, are commonly indicative of subaerial, arid, environments. Texture provides information about the 'energy' of the processes which have transported and deposited the sediment. For example, fast flowing 'high energy' rivers can transport large cobbles and boulders, while a slow moving 'low energy' river can only transport and deposit fine grained sands and muds. Composition is an indicator of the nature of environment. For example, shelly limestones are commonly found forming today

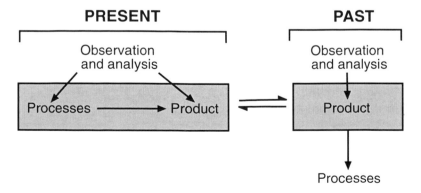

Figure 5.1 *Flow diagram illustrating the logic behind facies interpretation*

in warm, shallow marine environments, because shelly organisms thrive in such conditions. Similarly, sandstones rich in the mineral feldspar (arkosic sandstones) are indicative of an environment in which there has been little weathering, since feldspars are easily broken down into clay minerals by chemical weathering.

Sedimentary structures are extremely important in helping to deduce the processes by which different facies form. The most reliable structures are syndepositional structures, or those which formed as a direct result of the deposition of the sedimentary rock. They can provide direct evidence of whether the environment was one dominated by wind or water and can provide information on the direction of currents and the velocity of the fluid medium (wind or water) which deposited the sediment. For example, structures such as cross bedding (Figure 5.2) indicate the nature of the environment, the direction of current flow and are also useful way-up criteria.

Fossils were important in the initial recognition of the concept of facies and are still important in the recognition of facies in the stratigraphical record. On the crudest scale, through the basic application of actualism, fossils which closely resemble the marine organisms of the present day, such as corals, can be used to indicate ancient marine environments. In much the same way, study of the interaction of present-day organisms with their environment (i.e. ecology) can be used to interpret the relationship of ancient organisms with their environment (i.e palaeoecology). At the most detailed level, actual environmental parameters existing during facies formation can be calculated from the comparison of fossils with their nearest living relative, although with increasing geological age this becomes difficult. For example, some microfossils such as foraminifera and ostracods are environmentally sensitive today, and through direct comparison with their fossil relatives it is often possible to define precisely the prevailing conditions in the geological past.

Facies are defined on the basis of these four major characteristics and their environmental interpretation is developed through the study of modern sedimentary environments and products. Individual facies once recognised are named with regard to their product (e.g. sandstone facies), process (e.g. turbidite produced by a turbidity current) or environment (e.g. marine facies).

5.1.1 Facies Relationships

The relationship of one facies to another gives us the ability to assemble a picture of the geographical development of an area through geological time.

Sedimentary deposits do not continue infinitely in any one direction, but often grade laterally into other deposits or lithologies. This was recognised early in the development of geology. The use of fossils in determining relative chronology confirmed that adjacent but different sedimentary rocks were of the same age. It was apparent from Gressly's work that one adjacent facies might grade or interfinger (interdigitate) with another indicating the physical or geographical relationships between the original depositional environments, and underlining the importance of understanding the geometry of ancient rock bodies.

One concept which is of particular importance in understanding facies relationships was introduced by Johannes Walther in 1894. Walther studied recent sedimentary

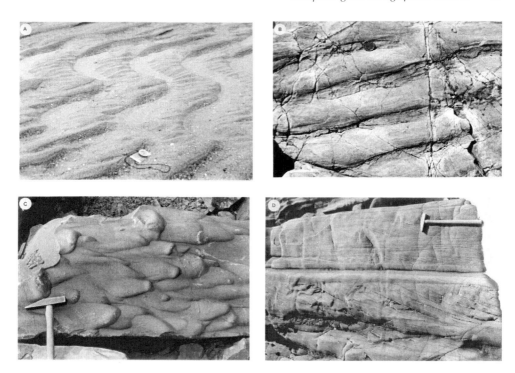

Figure 5.2   *Sedimentary structures: selected examples.* **A:** *Large ripples on the surface of a beach in south Devon, formed by water flowing from left to right [Photograph: M.R. Bennett].* **B:** *Large fossil ripples on a bedding plane within a series of fine grained sandstones [Photograph: D.G. Helm].* **C:** *Flute casts on the basal surface of a unit of Llandovery siltstone and sandstone from South Wales. As a turbidity current flows over a surface of sediments, turbulent flow within it produces a series of scour structures which are deeper up-flow and flare outwards in the down-flow direction. The subsequent deposition of sediment on top of the scoured surface produces a cast of the scour marks, like a plaster cast of a foot print. It is this cast which is shown in the photograph. Flute marks can be used to indicate the direction in which the turbidity current flowed, in this case from right to left [Photograph: BGS].* **D:** *Cross-bedding preserved in the Moine rocks of Morar, north-west Scotland (meta-sedimentary rocks). These structures were formed by a small sub-aqueous dune in a shallow marine environment. Sediment was moving from left to right. On such a dune sediment first moves up the gentle stoss side of the dune, some of this sediment is deposited in thin layers known as top sets, but most of the sediment avalanches over the steep lee face of the dune (down-flow) to form steeply dipping foreset beds. These foreset beds grade into bottom set beds, which approach the horizontal surface beneath the dune asymptotically. In the photograph one can see only foresets and bottom sets beds, the top sets have been removed by erosion, a common occurrence, as shown by the horizontal surface which truncates the top of the foresets. Cross-bedding like this can be used, therefore, to not only indicate the direction of sediment transport, but can also be used as way-up structures [Photograph: BGS]*

facies and their relationship to the environment in which they were deposited. From these observations he deduced that environments were not static through geological time. For example, rivers change their course, even on a human time scale, and consequently deposits produced by rivers should reflect such changes. Walther noted that as environments shift position in time, the respective sedimentary facies of adjacent environments will succeed each other in a vertical profile. Therefore, in sequences where there is no apparent break in the sedimentary record the vertical profile of sedimentary facies is equivalent to the lateral variation of facies at any one time. To put this crudely, if one was to turn a vertical profile on its side then it would give a picture of the lateral variation in the depositional environments present, during the period of time represented by the vertical profile. This principle has been used to interpret vertical profiles of sediments, such as one might encounter in an exposure or borehole, as the lateral variation of sedimentary facies present within an area at that time.

The concept can be illustrated by a hypothetical delta (Figure 5.3) modelled on the Mississippi Delta in North America. If one considers the sediments deposited at Point X for the three time intervals shown, one can see that at Time 1 pro-deltaic sediments are deposited followed by deltaic sediments as the delta progrades seawards. During Time 2 the delta has changed location due to the process of channel switching known as avulsion; Point X is no longer subject to deltaic, but marine sedimentation. Finally, during Time 3 the delta lobe has switched back to its original course and Point X is again receiving deltaic sediments. Consequently the vertical succession of sediments

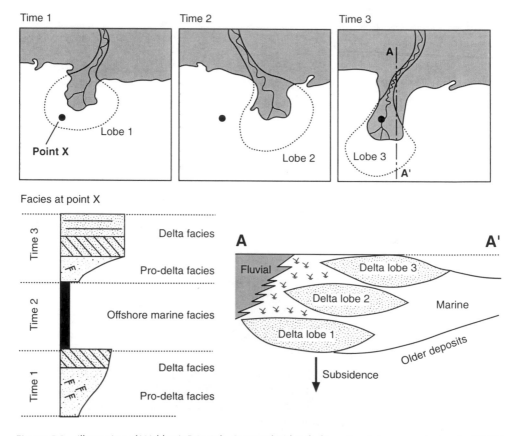

Figure 5.3 *Illustration of Walther's Principle. See text for details. [Based on information in: Leeder (1999)* Sedimentology and Sedimentary Basins, *Blackwell, Figures 22.10 and 22.11, p. 392]*

built-up at Point X during these three time intervals is a record of delta lobes switching location through time and the vertical log samples all the depositional environments present within the area at any given time.

Walther's principle is very valuable in situations where the relationship between facies is a purely sedimentological one, as opposed to a relationship caused by, for example, a rise or fall of sea level. Facies relationships which are purely sedimentological and are not the result of external variation (i.e. sea level) are known as autocyclic and usually reflect changes in the sedimentary environment. The gradual lateral movement of delta lobes as in our hypothetical example, is a response to increased sedimentary input and provides an example of an autocyclic relationship.

5.1.2 Relative Sea Level, Facies, and Sequence Stratigraphy

In the marine environment sea level has a profound effect on the spatial distribution and depositional environment of sedimentary facies.

Sea level has varied throughout geological time. The basic information that allows geologists to infer this is the pattern of marine, coastal and non-marine facies. Consequently in addition to the autocyclic mechanisms of the self-contained sedimentary system, controlled largely by sediment supply, the position and nature of facies may be controlled by external influences such as sea level or tectonic uplift. Such mechanisms are known as allocyclic. Often it is a difficult task to be able to recognise which (autocyclic versus allocyclic) has been of more importance in the development of a particular sedimentary sequence.

The relationship of facies through time can be used to deduce relative sea level variations from the stratigraphical record. The relationship of marine facies overlying non-marine or basement sequences shows that at some point sea level has risen to allow the sea to progressively transgress onshore. As a transgression proceeds, marine facies are said to onlap the land surface and it is the extent of this coastal onlap that allows the relative rise or fall of sea level to be deduced. In some sequences the progressive landward movement of marine facies means that successively younger sets of facies are seen to overlap the facies produced by the lower sea level stand, and lie directly onto the land surface beyond. In cases where the onlap is taking place over a folded, dissected basement, the term overstep is used. In this case the onlapping sedimentary facies are seen to step over the exposed, tilted strata of the older sequence. This, of course, leads to the production of an angular unconformity and it is now realised that most major unconformities are produced by such successive onlap (see Section 3.3.2).

Transgressive facies patterns are produced where successive landward onlap of marine sequences takes place during a rise in sea level. In a marine setting the different environments and their facies are usually in equilibrium, so that moving offshore we might encounter the following facies succession: a beach pebble facies, a nearshore sand facies, and an offshore mud facies. In a transgressive sequence these environments and their facies will simply move progressively inshore, building a vertical sequence which obeys Walther's principle. With falling sea level, a regressive facies pattern is produced with the same facies pattern moving progressively offshore. The

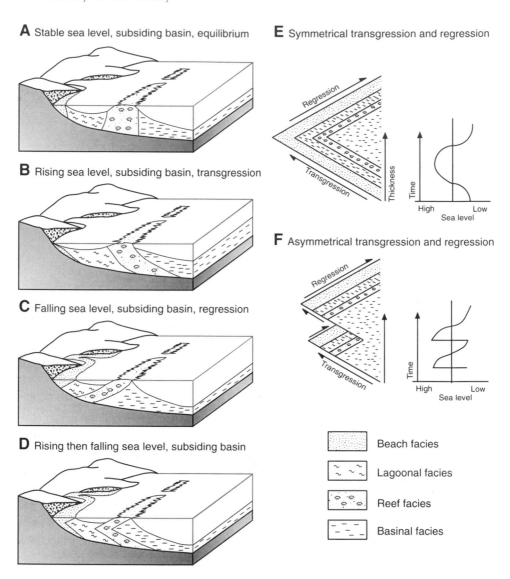

A Stable sea level, subsiding basin, equilibrium

B Rising sea level, subsiding basin, transgression

C Falling sea level, subsiding basin, regression

D Rising then falling sea level, subsiding basin

E Symmetrical transgression and regression

Regression

Transgression

High   Low
Sea level

F Asymmetrical transgression and regression

Regression

Transgression

High   Low
Sea level

Beach facies

Lagoonal facies

Reef facies

Basinal facies

Figure 5.4 *Facies patterns caused by the trangressions and regressions of the sea. **A–D:** Facies patterns associated with various sea level fluctuations within a subsiding basin. **E:** Stylised model of the facies pattern associated with a symmetrical transgression and regression of the sea. **F:** Stylised model of the facies pattern associated with an asymmetrical transgression and regression of the sea. Note the rapid trangression and associated lack of facies preservation*

movement of facies offshore is referred to as offlap. This concept is illustrated in Figure 5.4, which shows an offshore sequence of marine environments which consists of a beach, lagoon, reef, and basinal environments. In Figure 5.4A the facies patterns associated with a stable sea level, but gradually subsiding basin are shown; the facies do not shift laterally, but simply build upwards, maintaining equilibrium. In Figure

5.4B sea level rises, and although the basin subsides, the facies shift landward as the sea transgresses. Conversely if sea level was to fall the facies would shift seaward (Figure 5.4C). Fluctuations in sea level may, therefore, be recorded by a zig-zag pattern of facies shifts (Figure 5.4D). It is important to emphasise that this pattern of facies shifts will only be preserved if the basin is subsiding, thereby generating the space (accommodation space) in which the sediment can accumulate. It is also worth noting that in transgressive and regressive facies patterns the boundaries between individual facies are usually diachronous and that they cross time lines as they move progressively on or offshore. Figure 5.4E shows a stylised model of the facies patterns associated with a trangression and regression of similar duration. However, if the trangression is very rapid sedimentation may not occur sufficiently fast to record the shift and evidence of a trangressive or regressive episode may be absent (Figure 5.4F). In this way information can be deduced not only about the onshore and off-shore shift in facies, but also in some cases information about the rate of sea level change. Detailed information may, therefore, be obtained about sea level fluctuations using Walther's principle from a continuous record of interdigitated facies such as is seen in the Cenozoic sediments of south-east England (Box 5.2), or in the Late Mesozoic sediments of the western interior seaway of North America (Chapter 14).

Information about sea level change can also be obtained from a single log using Walther's principle, and can often be deduced in a simple fashion in the field. Figure 5.5 shows a student's field log of a coastal exposure through the Calcareous Sandstone Series (Lower Carboniferous) at Kinghorn on the Fife coast of Scotland. Each of the facies described are interpreted in terms of their relationship to sea level to deduce a crude idea of pattern of relative sea level change during this period.

The interpretation of individual facies and facies relationships provides, therefore, a set of tools with which patterns of sea level variation through geological time can be inferred. Sea level variation may involve either: (1) local or regional variations induced by local changes in land elevation (relative sea level) or (2) global or eustatic variations in sea level induced by changes in the volume of the world's ocean basins or by changes in the total volume of water present.

Determining the pattern of eustatic sea level variation through geological time is clearly of some importance. There are two methods by which a eustatic sea level curve can be deduced. The first method is based on estimates of the area of continental flooding, that is, the continental area covered by marine facies during a given period of geological time. Clearly the sea level rise needed to flood a given continent will depend on the range of land elevations present; a flat lying continent is easily flooded by a small sea level rise, while a continent with rugged relief will require a more substantial rise. The distribution of continental elevations on Earth is relatively constant, with a small area of extremely high and low elevations, and with most regions having elevations close to sea level. This distribution can be summarised in what is known as a hypsometric curve, a plot of the area present at each elevation. A hypsometric curve for the present globe is shown in Figure 5.6A. Assuming that this curve has been relatively constant through geological time it is possible to convert an area of continental flooding, as recorded by the distribution of marine facies on a continent, into an interpreted change in sea level. In order to achieve this, one must first plot the

Box 5.2
Facies patterns and sea level: an example from south-east England

The Palaeogene strata of the Hampshire and London basins in south-east England contain a remarkable record of sea level change. In these basins there was a competitive interplay between marine and non-marine environments. Stamp (1921) set up a simple model for a sea level cycle of transgression and regression which he called the 'cycle of sedimentation'. The facies which would result from this cycle were viewed as wedge-shaped incursions of marine deposits (Diagram A). Several sea level cycles would therefore produce a facies pattern which resembles a series of zig-zags or wedges. Stamp (1921) applied this simple model to the Eocene deposits of the Hampshire and London basins and summarised his results in a series of schematic diagrams such as the one shown below in diagram B. Stamp's work was refined by Curry (1965), although the same zig-zag pattern of facies was identified (Diagram C). The sea level changes recorded in south-east Britain are probably eustatic in nature and may be explained in terms of variations in the rate of sea floor spreading in the opening of the Atlantic.

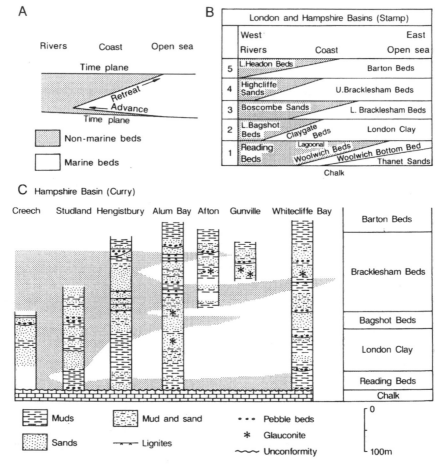

Sources: Stamp, D. 1921. On cycles of sedimentation in the Eocene strata of the Anglo–Franco–Belgian Basin. *Geological Magazine* **108**, 108–114, 146–157, 194–200. Curry, D. 1965. The Palaeogene beds of south-east England. *Proceedings of the Geologists' Association* **76**, 151–173. [Diagram modified from: Stamp (1921) *Geological Magazine* **108**, Figure 1, p. 108; Curry (1965) *Proceedings of the Geological Association* **76**, Figure 3, p. 167.]

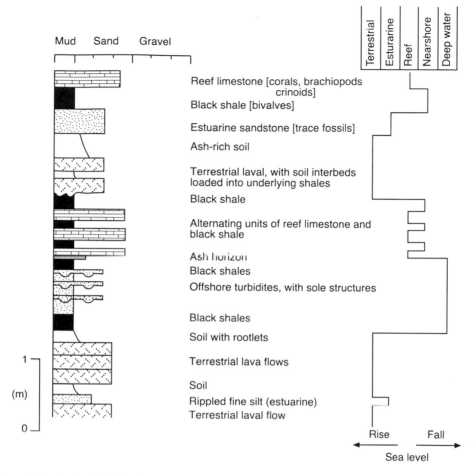

Figure 5.5   *Student field log through part of the Calcareous Sandstone Series (Lower Carboniferous) at Kinghorn on the Fife coast of Scotland. Note how changes in relative sea level can be deduced from the facies logged*

distribution of marine facies of a given age on a palaeogeographical map of similar age and then measure its area (i.e. the area of continental flooding). This area can then be converted into a sea level height using a continental hypsometric curve as shown in Figure 5.6B. This operation is then repeated for successive time intervals in order to reconstruct a sea level curve (Figure 5.6C). This type of approach has been used by Hallam to produce a eustatic sea level curve for the Phanerozoic (Figure 5.6D). He concentrated on the North American and Eurasian continents and assumed that the hypsometric curve in the past was similar to that of today. In fact, this approach has been modified by recent work which has suggested that the shape of the hypsometric curve varies with land area; large continental areas have higher curves than smaller continents. Since the number and size of continents has changed through time so has the hypsometric curve. Despite these recent modifications, the essential trends remain unchanged (Figure 5.6D).

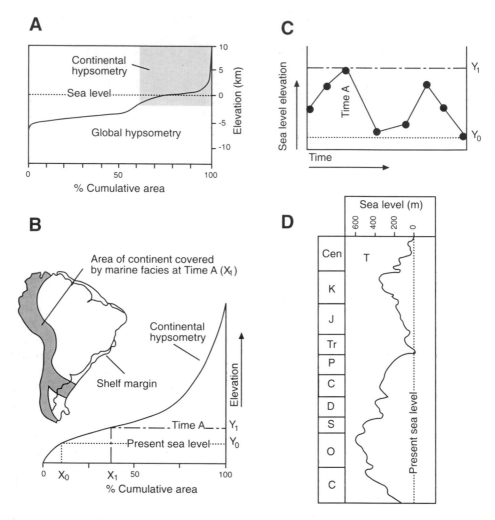

Figure 5.6 *Methods used by Hallam and others to reconstruct eustatic sea level variations from the area of continental flooding.* **A:** *A global hypsometric curve.* **B:** *Method for determining changes in sea level elevation from the area covered by marine facies at a given time.* **C:** *Construction of a sea level curve by repetition of the method outlined in (B) for different time intervals.* **D:** *The eustatic sea level curve constructed by Hallam using these methods.* [Modified from: **A–C:** Hallam (1998) In: Doyle and Bennett (Eds) Unlocking the Stratigraphical Record, John Wiley & Sons, Figure 15.7, p. 432; **D:** Hallam (1984) Annual Reviews of Earth and Planetary Science 12, Figure 5A, p. 220]

The second method involves the use of relative onlap or offlap. Stable continental margins contain a remarkably detailed sedimentary record of sea level variation. However, these sediments can only be analysed by seismic exploration, which employs a technique similar to echo sounding. If a sound signal is sent towards the sea floor and the time taken for it to return is recorded (i.e. the echo time) then the depth of water can be determined if the velocity of sound in water is known. If a more powerful source of energy with a lower frequency is used then the energy will penetrate the sea floor and be reflected from the boundaries between sediment layers. These boundaries are commonly referred to as seismic reflectors, or just reflectors for short. A seismic profile of the sediments beneath the sea floor can be obtained from seismic exploration. Such a profile resembles a geological cross section, although the horizontal dimension records the ship's movement and the vertical scale records echo time rather than depth. The reflectors are commonly spaced somewhere in the region of 10 metres apart and are therefore unlikely to be truly representative of the full stratigraphical sequence. However, when used in conjunction with boreholes and sediment cores a detailed picture of the rocks and sediments which underlie a section of sea floor can be obtained. Most of the reflectors which return echoes, and are therefore recorded in seismic profiles, represent surfaces which were once the sea floor. As a consequence they possess a full suite of onshore-offshore facies along their length and are effectively facies-independent time-planes. Each former sea floor and the sediment associated with it may either onlap or offlap previous surfaces or sea floors (Figure 5.7). By determining the pattern of relative onlap or offlap one can obtain a record of the marine transgressions and regressions which have taken place in the past.

Seismic reflection studies of continental margins have shown that the pattern of onlapping and offlapping surfaces can be grouped into sediment packages, known as sequences. Each sequence is usually bounded, top and bottom, by an unconformity. The onlapping or offlapping units within a sequence are known as parasequences. Stable continental margins often possess a stack or succession of sequences produced by the transgression and regression of the sea. Sequences develop at the continental margin as it is here that the major sedimentary bodies, such as deltas, form. This contrasts with the centre of the basin where sedimentation only occurs by the settling out of fine grained sediment from suspension and there is little input of coarse sediment. Given a fall in sea level, a delta will build out seawards, or prograde, into the available space (the accommodation) of the sedimentary basin. Deltas and other major sedimentary systems are in equilibrium with sea level: that is they change their location in sympathy with changes in sea level.

The concept of sediment packages or sequences along continental margins is illustrated in the block diagrams in Figure 5.8. These diagrams depict the pattern of sediment packages associated with a hypothetical continental margin subject to sea level fluctuations. In the first block diagram an initial wedge of sediment is seen to build up during a period of high sea level (Figure 5.8A). In the second block diagram a fall in sea level causes the erosion and truncation of this sediment wedge and the deposition of a sediment fan offshore (Figure 5.8B). The initial sedimentary wedge is truncated by erosion and an unconformity is formed which acts as an upper boundary to the initial, high sea level sediment wedge. Sea level has now stabilised at a low point in the third block diagram and a new sediment package builds up. Finally as sea level rises to its former level a new sediment package accumulates on top as a delta, or similar sediment body, builds up (Figure 5.8E). When viewed in an interpreted seismic section the sediment packages along this coastal margin should resemble those in Figure 5.8F. Two sediment sequences can be recognised separated by the unconformity formed during the sea level low.

We can explore the concept of sequence development further with reference to the hypothetical continental margin seismic profile which we have interpreted in Figure

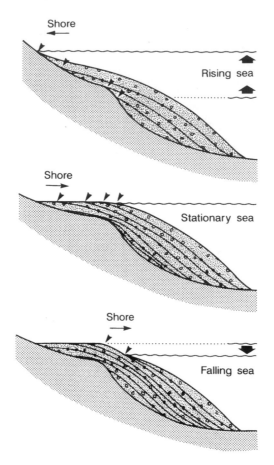

Figure 5.7 *Onlap and offlap of sedimentary bodies. The upper illustration shows how rising sea level produces a pattern of onlapping sedimentary bodies, in which each unit, indicated by the arrows, extends further onshore than the last. The middle diagram shows what happens when sea level is stationary, in which each unit is displaced seawards (offlapping) by the building out of a sedimentary body. The lower illustration shows how a fall in sea level also produces a pattern of offlapping sedimentary bodies, in which the extent of each unit is further offshore than the last. [Modified from: Van Andel (1985)* New Views on an Old Planet, *Cambridge University Press, Figure 10.4, p. 152]*

5.9. Figure 5.9A is an idealised section through a basin margin and shows the pattern of former sea floors (seismic reflectors) as well as the pattern of facies. Notice that the facies boundaries are diachronous and along each time plane or former sea floor the same succession of facies is present. Figure 5.9B shows the main unconformities or sequence boundaries present within the section. Each of these unconformities was formed by a fall in sea level. These boundaries define three sequences. Figure 5.9C shows both the sequence boundaries and the former sea floor surfaces which split each sequence into parasequences. The relative coastal onlap and offlap of the para-sequences can be seen clearly in this section. The history of this margin can be reconstructed as follows (Figure 5.9C).

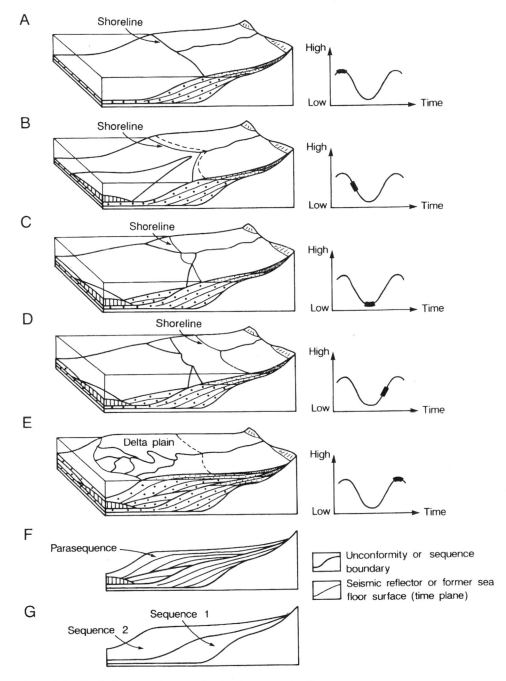

Figure 5.8 *Block diagrams showing the development of sedimentary packages or sequences along a continental margin with fluctuating sea level. The graphs on the right hand side of the diagram show the position of sea level within a simple cycle of sea level change from high, through low and back to high. [Modified from: Posamentier et al (1988). In: Wilgus et al (Eds) Sea Level Change: an Integrated Approach, Society of Economic Palaeontologists and Mineralogists, Special Publication 42, Figures 1–6, pp. 111–114]*

1. Parasequences 1 to 3 were deposited during a rise in sea level, the magnitude of which is illustrated by the amount of coastal onlap. Sea level reached its maximum elevation during parasequence 3. Parasequences 4 to 6 were deposited as sea level began to slowly fall, showing an offlapping relationship, which is partially obscured because of the later truncation at the sequence boundary.
2. Following the deposition of parasequence 6, sea level fell rapidly. Erosion of parasequences 1 to 6 produced an unconformity and the first sequence boundary. The eroded material was deposited as a small offshore fan, parasequences 7 to 8.

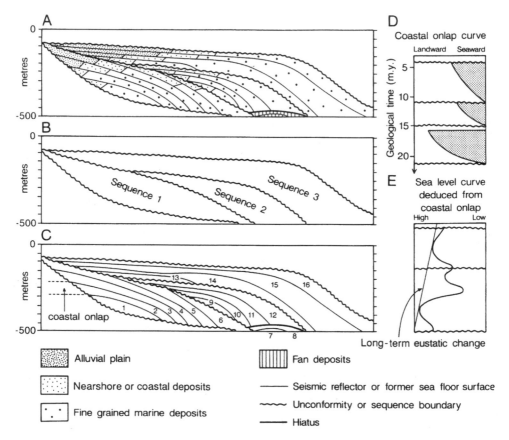

Figure 5.9 *Hypothetical example of an interpreted seismic profile across a continental margin.* **A:** *Shows an interpreted seismic profile, showing both the facies pattern and the seismic reflectors (former sea floors) within the section.* **B:** *Within this section the sequence boundaries have been picked out, and three sedimentary sequences have been identified.* **C:** *Shows the parasequences within each of the three identified sequences.* **D:** *Shows the coastal onlap curve produced by the study of the onlap and offlap patterns within the seismic profile (C).* **E:** *Illustrates how the coastal onlap curve can be combined with a knowledge of long term eustatic patterns to produce a sea level curve. [Modified from: Wilson (1992). In: Brown et al (Eds)* Understanding the Earth, *Cambridge University Press, Figure 20.18, p. 411]*

3. A subsequent rise in sea level deposited parasequences 9 to 12 and again the amount of sea level rise is indicated by coastal onlap onto sequence boundary 1.

4. Before sea level had risen to its former level it fell rapidly again causing erosion of parasequences 7 to 12. This produced a second erosional surface and sequence boundary.

5. Sea level then rose dramatically to deposit parasequence 13. Further sea level rise and progradation of the sediment bodies occurred during the deposition of parasequences 13 to 16.

This hypothetical example illustrates the principle by which sequences can be recognised and interpreted in terms of sea level variation. Three sequences can be recognised, each with bounding unconformities. These unconformities are produced by the action of relative sea level. By carefully measuring the amount of coastal onlap within each sequence a coastal onlap curve can be produced (Figure 5.9D) from which a sea level curve can be deduced given a correction for any change in land elevation (Figure 5.9E).

It is important to note that sequences are easily recognised only at basin margins. Offshore, sequence boundaries (unconformities) may equate with conformable units with no recognisable erosional surface (conformities) since sedimentation within the basin centre may have been continuous despite the fluctuation at the basin margin (Figure 5.10).

The study of the interrelationships of these sediment wedges or packages therefore reveals a history of local sea level change which can be obtained for any given continental margin. Correlation of local sea level histories obtained from continental margins spread across the globe allows a picture of eustatic sea level to be obtained. This type of technique has been used to obtain a eustatic sea level curve commonly referred to as the Vail sea level curve after the principal researcher in this field (Figure 5.11).

Consequently, it is possible to obtain a eustatic sea level curve either using the area of continental flooding (Hallam) or by using sequence stratigraphy (Vail). In outline the Vail and Hallam eustatic sea level curves are very similar despite the fact that they are obtained by different methods (Figure 5.12).

The study of sea level change through sedimentary sequences was first developed by oil companies as a by-product of their exploration for offshore oil reserves. More recently, however, the concept of sedimentary sequences has been used as a tool in correlation both within a depositional basin and between basins (see Section 7.1.2). This application is known as sequence stratigraphy and relies simply on the recognition and correlation of depositional sequences and the unconformities which define them. Most sequence stratigraphers argue that sequences are produced in response to allocyclic mechanisms, and primarily to changes in relative sea level as we have discussed above. The bounding unconformities of sequences are inferred to be genetically linked to changes in sea level. If the change in sea level is eustatic (worldwide) rather than local, then it follows that the resulting sedimentary packages can be correlated on the basis of sequence boundaries from one continental margin to another. Some researchers have, however, questioned whether relative uplift of continental areas or even increased sediment supply (an

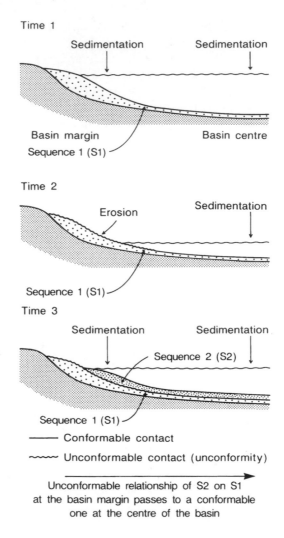

Figure 5.10 *Sequence boundaries and their conformity in the basin centre. This diagram illustrates how the boundaries between sequences are usually unconformities at the basin margin, becoming conformities in the basin centre.* **Time 1:** *deposition of sequence 1.* **Time 2:** *a sea level fall exposes sequence 1 to erosion at the basin margin, although sedimentation is continuous in the centre of the basin.* **Time 3:** *a rise in sea level results in the deposition of a second sequence (sequence 2) which rests at the basin margin on the eroded surface of sequence 1 forming an unconformity. In the centre of the basin, sequence 2 rests conformably on sequence 1 as sedimentation has continued uninterrupted*

autocyclic mechanism) could not have caused the patterns of relative onlap and offlap seen within many sequences. Whatever the cause of sequence boundaries (allocyclic verses autocyclic) sequence stratigraphers have had some success in the correlation of sedimentary sequences and the unconformities which bound them.

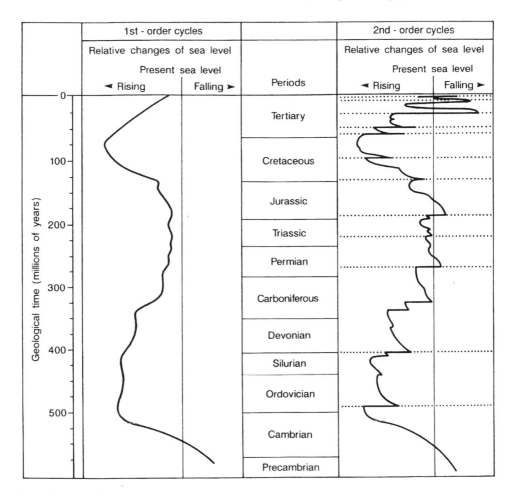

Figure 5.11 *Vail sea level curve produced from the analysis of sedimentary sequences within seismic sections. [Modified from: Vail et al (1977)* American Association of Petroleum Geologists, Memoir 26, *Figure 1, p. 84]*

5.2 Palaeogeography

Palaeogeography is the study of the geography of the ancient Earth. On the broadest scale palaeogeography refers to the distribution of the continents and oceans, and one of the most important tools in interpreting the positions of ancient continents has been palaeomagnetism. As discussed in Section 4.2.1, when igneous rocks cool in a given continent, their crystals preserve a record of the position of the continent relative to the location of the poles, which although subject to small variations, are relatively constant through geological time. In this way the motion of the continents across the globe is recorded in the palaeomagnetic record locked within their igneous rock record. Three types of information are required to achieve this:

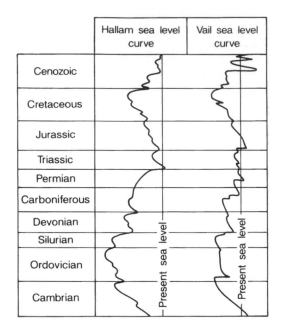

Figure 5.12   *Comparison of the Vail and Hallam sea level curves. [Modified from: Hallam (1984)* Annual Reviews of Earth and Planetary Science *12, Figure 5, p. 220]*

1. The orientation relative to the poles, determined by the declination, which is the angle between the geographic and magnetic poles.
2. The latitude, determined by the inclination, which is the angle that would be taken by a freely suspended magnetised needle relative to the earth's magnetic field; this varies according to latitude, from vertical at the poles to near horizontal at the equator.
3. The longitude, which can not be determined from palaeomagnetism.

From this discussion it is clear that palaeomagnetism can be used to provide a palaeolatitude, but longitude has to be determined from other geological data. Such data may rely upon the fit of continental blocks, and the location of faunal elements around the margins of the continents, which becomes increasingly difficult in the more distant geological past. It is for this reason that reconstructing Early Palaeozoic and Precambrian continental positions is ambiguous and a fact which is reflected in the increasing number of continental positions for conflicting reconstructions (Box 12.2).

Smaller scale geographies for a given area may be obtained through the study of facies distributions. These may ultimately be gathered together to give a detailed pattern of the geography of an area like Europe or North America through time. Local palaeogeographical maps are constructed by examining the distribution of facies of a similar age within a given area, and plotting the relative position of the different environments indicated by the facies. Such a palaeogeographical map might show the distribution of marine, coastal and continental environments.

Box 5.3
The palaeogeography of a submarine fan

Kleverlaan (1989) provides a case study in production of a palaeogeographical map, in this case of a submarine fan in the Tabernas Basin of south-east Spain. The Tabernas Basin is one of several Neogene basins within the Betic Cordillera. This Miocene basin was formed by tectonic deformation in the Serravallian and was infilled in the Tortonian by continental and then marine sediments. The northern margin of the basin at this time consisted of a steep cliffed coast fringed by coral reefs, while a series of submarine fans infilled the basin floor. One of these was the Tabernas fan, the palaeogeography of which was reconstructed by Kleverlaan (1989). This fan is composed of a range of turbidite and other mass flow deposits associated with the downslope movement of sediment from the basin margins. Due to the exceptional level of exposure Kleverlaan (1989) was able to trace the geometry of this fan and associated lobes and feeder channels during a single time slice. He used an event horizon, the Gordo Megabed, which forms a prominent marker horizon within the basin to help define his time slice. This unit was formed by a widespread seismically induced mass flow within the basin. By careful mapping and logging of the fan sediments below this marker horizon Kleverlaan (1989) was able to reconstruct a palaeogeographical map of the Tabernas fan during a specific time interval. By careful observation he built up a three-dimensional picture of the sediments present during his chosen time slice, from which he was able to construct a map of the ancient sea floor. This example illustrates how careful field investigations in an area of excellent geological exposure can be exploited to produce a palaeogeographical map, in this case of a former sea bed.

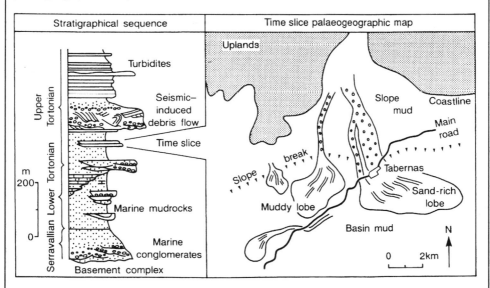

Sources: Kleverlaan, K. 1989. Three distinctive feeder-lobe systems within one time slice of the Tortonian Tabernas fan SE Spain. *Sedimentology* **36**, 25–45. [Modified from: Kleverlaan (1989), *Sedimentology* **36**, Figures 2 and 6, pp. 26 and 27]

Palaeogeographical maps are constructed by simple facies interpretation carried out for specific snapshots of geological time known as time slices (Box 5.3). A time slice is determined using biostratigraphy or other correlation tools which are facies independent. Once the time constraints are determined, stratigraphical sections can be correlated within a given area and the spatial pattern of facies within it at a given time can be determined. A palaeogeographical map is then constructed by interpreting each facies, so that a clear understanding of the nature and position of environments within the area is achieved. If this procedure is repeated for earlier and later time intervals then a series of palaeogeographical maps can be compared to show how an area or region has changed through time. These general principles are illustrated in Figure 5.13. In this example logs are taken at six locations, including coastal cliffs, river cliffs and a bore-hole, and the information plotted on a chronostratigraphical log (i.e. using time as the vertical axis). This is achieved by first carefully recording the facies present, and then by using an independent dating method, such as biostratigraphy or event stratigraphy, each unit is correlated in time in order to provide a succession of chronostratigraphical units. Time slices can then be established, and the facies deposited at each point for each time slice can be plotted on a map. By careful interpolation a palaeogeographical map for this period (Time X) can be fleshed out, providing an approximation of con-temporary environmental conditions (Figure 5.13).

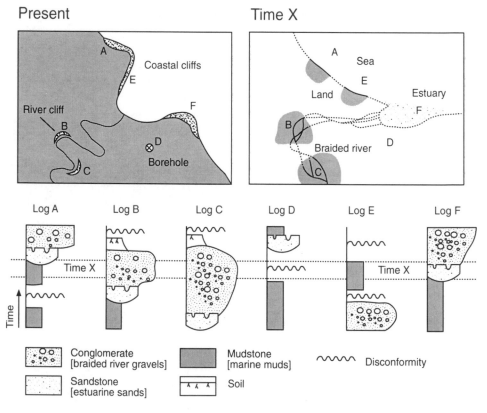

Figure 5.13   *Hypothetical illustration of the methodology involved in the construction of a palaeogeo-graphical map. See text for details*

5.3   Palaeoclimatology

The goal of palaeoclimatic reconstruction is to define the climatic conditions and characteristics of the atmosphere, oceans and landmasses through geological time. The challenge is to find geological evidence that can be used as the thermometers, barometers and anemometers of different geological periods. There are four main types of data available to reconstruct palaeoclimate: (1) chemical data; (2) biological data; (3) physical data; and (4) computer simulations.

5.3.1   Reconstructions Based on Chemical Data

The most widely used chemical method in climatic reconstruction is the measurement of stable isotopes (Box 4.4). In particular, the oxygen isotopic composition of carbonates, secreted by organisms growing in sea water can be used to provide palaeotemperature data. On average the ratio of ^{18}O to ^{16}O in today's oceans is about 1 to 500 or alternatively 0.2 per cent of all oxygen is ^{18}O. This ratio is highly sensitive to temperature (Figure 5.14). Consequently, if the oxygen isotope ratio present in the calcium carbonate ($CaCO_3$) of a shell is known, an indication of the ocean temperature in which it lived can be obtained. Planktonic organisms known as foraminifera are the most frequently used organisms for palaeo-temperature analysis because of their widespread distribution over the surface of present and former

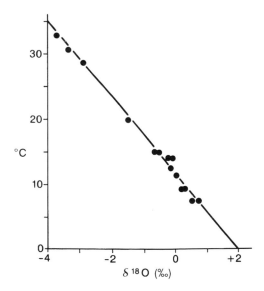

Figure 5.14   *Graph to show the effect of temperature on the proportion of ^{18}O. This graph is based on laboratory experiments conducted on the oxygen isotopic composition of molluscs growing at different temperatures and demonstrates that oxygen isotopes in carbonate shells can be used as palaeotherm-ometers. The oxygen notation expresses the difference in the abundance of ^{16}O and ^{18}O from a standard in parts per thousand. [Modified from: Barron (1992). In: Brown et al (Eds) Understanding the Earth, Cambridge University Press, Figure 24.1, p. 487]*

oceans. However, a wide range of other fossil organisms with calcareous shells can also be used.

Four factors limit the use of oxygen isotopes as a palaeoclimatic tool. First, the isotopic composition of sea water is dependent on evaporation, precipitation and run-off from land areas. Evaporation selectively removes the 'lighter' ^{16}O isotope (as 'light' water: $H_2^{16}O$) and correspondingly rain and snow is enriched in ^{16}O. If this rain or snow returns to the oceans the oxygen isotope ratio of the oceans remains unchanged, but during a glaciation, rain and snow is stored in ice sheets and does not return quickly to the oceans. Consequently, during glaciation oceans are enriched in ^{18}O (Figure 5.15). This is very useful in the study of the Quaternary 'Ice Age', since the changing ratio of ^{18}O to ^{16}O in the skeletons and shells of such organisms as foraminifera taken from ocean sediments can be used to provide a record of the growth and decay of ice sheets (Box 5.4). However, if one wants to obtain palaeo-temperature estimates a correction for the presence of ice sheets is required and therefore a knowledge of changes in the global ice volume. This can provide a serious limitation when considering the pre-Quaternary record, where our knowledge of the Earth's glacial history is incomplete.

The second limiting factor is that not all organisms live in balance with the isotopic composition of sea water. For example, the shells of certain organisms may have isotopic compositions which are different from that of the sea water in which they live and do not therefore record changes within it.

Third, the use of isotopic data requires a knowledge of the geographical habitat of the species being used in the analysis. For example, if the organism was mobile then this could seriously affect the interpretation of the results, the palaeotemperature estimates being an integration of the different locations or environments in which the organism lived. Finally, the isotopic composition of a shell may change during the

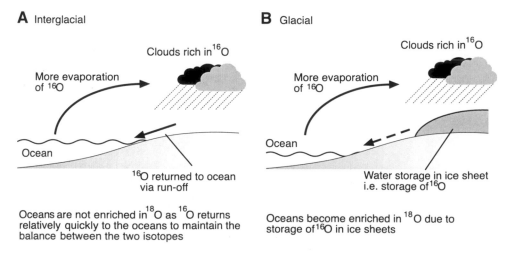

Figure 5.15   *Diagram to show the natural fractionation process which takes place during the evapora-tion of sea water and illustrates how the growth of an ice sheet disrupts the balance of oxygen isotopes within sea water*

Box 5.4
Glacial history and the oxygen isotope record

The isotopic composition of ocean water changes with the growth and decay of ice sheets. A record of this isotopic variation and therefore a record of global ice volume is stored in the carbonate of shells within deep sea sediments. The ratio of ^{18}O to ^{16}O in today's oceans is normally about 1:500 and the oxygen locked up in the carbonate of marine shells reflects this ratio. This ratio varies with the growth and decay of the Earth's ice sheets. When sea water evaporates a process of natural fractionation occurs and more water molecules with ^{16}O are evaporated than those with ^{18}O because the atomic mass of ^{16}O is less (i.e. the molecule is 'lighter'). Atmospheric water, clouds and rain, are therefore enriched in ^{16}O. In a non-glacial environment the balance of ^{18}O to ^{16}O is maintained because rainwater falling on land quickly returns to the ocean via rivers. In contrast, during a glacial period the oceanic balance of ^{18}O to ^{16}O is upset because atmospheric moisture is not returned quickly to the oceans but falls as snow and is stored in ice sheets (Figure 5.15). Consequently the oceans are enriched in ^{18}O during a Glacial Period. A record of the Earth's glacial history during the Quaternary 'Ice Age' can, therefore, be obtained from an analysis of oxygen isotopic composition of deep ocean sediments: horizons in which the shells and skeletons of marine organisms are enriched in ^{18}O correspond to period when the Earth's total ice volume was large (Glacial Periods) while those horizons with less ^{18}O correspond to periods with a small global ice volume (Interglacial Periods).

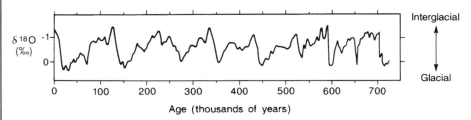

Sources: Imbrie, J. and Imbrie, K.P. 1979. *Ice Ages: Solving the Mystery*. Harvard University Press, Cambridge, Massachusetts. [Diagram modified from: Barron (1992). In: Brown et al (Eds) *Understanding the Earth*. Cambridge University Press, Figure 24.2, p. 487.]

transformation of a sediment to a rock (diagenesis) and may not therefore reflect the original composition of the ocean in which the shell lived. Despite these limitations the analysis of oxygen isotopes can provide important palaeoclimatic data.

The isotopes of carbon (^{12}C, ^{13}C, ^{14}C) contained within the shell of marine organisms can also be used to give an indication of the organic productivity of the oceans and the pattern of ocean circulation. This information is very important given the close coupling of the oceanic and climatic systems.

5.3.2 Reconstructions Based on Biological Data

Palaeoclimatic information can be obtained from fossils. If a certain plant or animal only lives in a specific climatic environment today then its distribution can be used to provide information on the extent of that environment, and provide an interpretation of the past. This principle is well-illustrated by large reptiles or amphibians which are almost exclusively tropical today. For example, crocodiles are not found poleward of latitudes with a mean annual temperature of 15°C. Consequently the presence of fossil crocodile bones within the geological record can be used to infer temperatures in excess of 15°C. In the same way hermatypic corals, which are the principal constituent of modern reefs such as the Great Barrier Reef in Australia, are largely restricted to the tropical and subtropical latitudes; less than 30° N or S and in waters warmer than about 21°C (see Section 5.4 and Figure 5.19). Reef limestones within the geological record which contain hermatypic corals provide, therefore, a good indication of surface water temperature.

One of the major difficulties with using a given species of plant or animal as a palaeoclimatic indicator is that most organisms can survive in a range of climatic conditions. This may lead to large margins of error in any reconstruction of palaeo-climate. To avoid this problem an assemblage of fauna and flora is usually used. The overlap between the climatic preferences of one species and another helps reduce the range of possible palaeoclimates which can be inferred (Figure 5.16). This method is often referred to as the method of mutual climatic ranges. The application of this type of approach is well-illustrated by its successful use in reconstructing the palaeo-temperature of the last 22 000 years of the Quaternary from beetle remains (Box 5.5).

Plants are particularly important providers of palaeoclimatic information. It is well-known that many plants are restricted by climatic type, and that there is a broad correspondence between floral zones and climatic belts today. The identifica-tion of latitudinal patterns in fossil plants has therefore been used as an indicator of the ancient climatic patterns. Detailed aspects of plant morphology are also impor-tant. For example, plants which live in arid areas have reduced leaf areas or have leaves which turn end-on to the sun to avoid excessive water loss through evapotran-spiration, while plants typical of a rain forest tend to possess elongated leaf ends or 'drip tips' which are adapted for shedding water. By way of contrast, leaves in tem-perate forests are much more variable in form, often with crenulations at their mar-gins. This difference in leaf morphology is widely used in palaeoclimatology, since the proportion of plant species within a community with an 'entire margin' (i.e. without

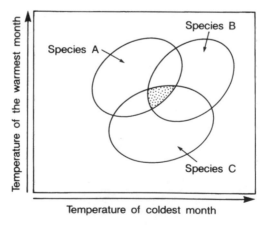

Figure 5.16   *This schematic diagram illustrates the concept of mutual climatic ranges. Three species (A, B and C) each have broad climatic preferences. A rock containing all three species as fossils is likely to have formed in a climate corresponding to the stippled area*

Box 5.5
The mutual climatic range method and the reconstruction of palaeo-temperatures during the last 22 000 years

Beetle or coleopteran remains can provide good palaeo-temperature data. As a group beetles are varied and many species are climate sensitive with well-defined tolerance ranges. They are found in a wide variety of Quaternary sediments and most of the species recorded in these sediments exist today.

Atkinson et al (1987) used Coleoptera to reconstruct the palaeo-temperatures of Britain over the last 22 000 years. These palaeo-temperature reconstructions were made by using the technique of mutual climatic ranges. The basic assumption is that, if the present day climatic tolerance range of a beetle is known, then fossil occurrences of that species imply a palaeoclimate which was within the same tolerance range. If several species occur together as a fossil assemblage, then the palaeoclimate at the time they were alive must lie within the mutual intersection of their tolerance ranges. This intersection will be smaller and the palaeoclimate deduced more precise if a large number of co-existing species are used. Atkinson et al (1987) first defined the climatic tolerance range of 350 living beetle species which commonly occur as Quaternary fossils. This was done by compiling a map of each species' present distribution and defining it in terms of temperature variables. The second phase of this exercise involved the selection of dated fossil beetle assemblages from 25 different locations within the British Isles. The mutual climatic range of each of these assemblages was determined by superimposing the climatic range of each species present. A range of possible palaeo-temperatures was thereby deduced for each assemblage and the median value calculated. By plotting the age of each assemblage against the deduced palaeotemperature a graph of temperature variation may be obtained as illustrated below.

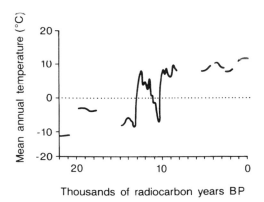

Source: Atkinson, T.C., Briffa, K.R. and Coope, G.R. 1987. Seasonal temperatures in Britain during the past 22 000 years, reconstructed using beetle remains. *Nature* **325**, 587–592. [Diagram modified from: Atkinson et al (1987) *Nature* **325**, Figure 3, p. 592.]

crenulations) is closely correlated with mean annual temperature (Figure 5.17). The presence and nature of tree rings in fossil wood are also useful in palaeoclimatology, as pronounced rings indicate seasonality, with a definite growing season, while subdued rings indicate continuous growth. For example, fossil wood from Cretaceous high latitudes such as Antarctica show these marked seasons in an otherwise warm globe devoid of extensive ice.

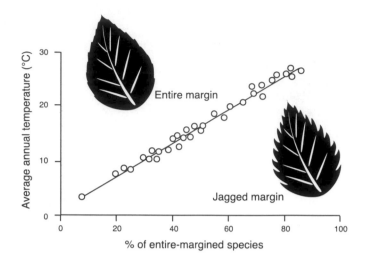

Figure 5.17   *Relationship between climate and leaf shape. As plants become more ecologically stressed they tend to have more irregular leaf margins in order to reduce water loss via evapotranspiration. [Modified from: Stanley, S.M. (1989)* Earth and Life Through Time, *Freeman, Figure 2.21, p. 43]*

In general, for a fossil to be of use as a palaeoclimatic tool it must meet the following criteria. First, it must live in equilibrium with its environment, so that any changes to that environment will lead to either adaptation, migration or mortality. Second, the principle of uniformitarianism must apply. The climatic preferences of species today must have been constant throughout their existence if they are to be used as indicators of past environments. This becomes difficult for fossil groups without living relatives. In this case the closest living relative to a fossil in either genetic or morphological terms is often used: this can, however, be misleading.

5.3.3   Reconstructions Based on Physical Data

Sedimentary rocks can be used to provide palaeoclimatic information, if the environment in which they were deposited was climatically controlled. The distribution through time of climate-sensitive deposits, like coals, evaporites, tillites and carbonates, have in the past been zonally distributed in much the same way as they are today (Figure 5.18), illustrating their potential importance in palaeoclimatic reconstruction. Three broad types of palaeoclimatic indicators can be recognised: (1) those indicative of warm arid climates; (2) those indicative of wet or humid climates; (3) those which indicate polar or glacial conditions.

1.   Lithologies and deposits indicative of warm or arid climates.   Within the geological record evaporites are probably the best indicators of global aridity. Today they are precipitated on or below the surface of the soil in arid or semi-arid regions and include rock salt and gypsum. Evaporites are characteristic of rainfall deficient, subtropical

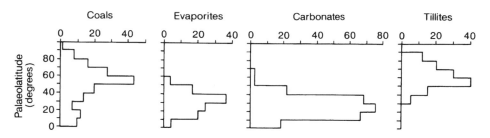

Figure 5.18 *Palaeolatitudinal zonation of climatically sensitive deposits. Each histogram shows the frequency of deposits against palaeolatitude. Evaporites and carbonates are concentrated in the low latitudes, while tillites and coals are mostly found in higher latitudes. [Modified from: Scotese & Summerhayes (1986) Geobyte 1, Figure 1, p. 29]*

belts with high rates of evaporation. Their occurrence within the geological record provides, therefore, a good indication of arid conditions.

Sediments with a strong red coloration, which is usually caused by the precipitation of iron oxide, are also sometimes taken as indicative of arid climates, although red beds can also occur as a result of fluvial deposition in a variety of environments. The presence of feldspar-rich (arkosic) sandstones has also been used as an indicator of arid climates, but is again somewhat unreliable, since arkosic sandstones can form in a variety of environments.

When sand dunes, in either coastal or desert regions, become lithified into sandstone a record of the wind regime in which deposition occurred may be recorded. The internal structure of aeolian sandstones can be difficult to interpret but are commonly used to give indications of palaeo-wind directions. Extensive deposits of aeolian sandstone may also be used as indicators of desert-like environments.

The deposition of calcium carbonate as limestones or chalks may also be used to indicate a warm climate. The solubility of calcium carbonate decreases as temperature increases and therefore high temperatures tend to lead to more limestone. Care is, however, required since the presence of abundant limestone is also correlated with the periods of high sea level.

2. Lithologies indicative of wet or humid climates.
Both bauxite and coal can be used as indicators of wet or humid climates. Bauxite is the end product of intense chemical weathering and soil leaching, typical of a tropical location. It develops from weathering of rocks rich in aluminium, at sites which experience intense leaching and where rainfall exceeds 1200 to 1500 mm a year. Its palaeoclimatic significance lies in its association with high rainfall. The presence of coal can also be used to indicate wet or humid conditions. Coal forms from the deposition of vegetation in anoxic conditions where organic material does not break down and decay: today's peat bogs may be tomorrow's coal. Peat bogs generally occur in high-latitudes with damp humid climates. Woody material does not normally accumulate in lower latitudes due to the higher temperature which results in rapid oxidation and breakdown by bacteria. There are however complications, since the principal prerequisite for coal – poor drainage – may occur even in semi-arid areas given a high water table.

Consequently, there are limitations to the use of peat and coal as temperature and precipitation indices, but with care they can sometimes be used.

Fossil soils (palaeosols) and weathered layers may also give some indication of palaeoclimate. The characteristics of a soil are partly determined by climate while the intensity of a weathered profile and the products of that weathering are also climatically controlled. Therefore the presence of weathered horizons or soils within the geological record may provide valuable clues as to the former climate.

3. Lithologies indicative of cold or glacial climates.   There are two main indicators of cold or glacial climates. The first of these is the presence of tillites. Glaciers and ice sheets deposit glacial sediment known as till or boulder clay, which when lithified are known as tillites. Tillites provide evidence of glaciation. Modern till usually consists of large sub-rounded pebbles or boulders set in clay or silt. The pebbles or boulders present tend to be ice scratched (striated) and aligned in the direction in which the glacier flowed while the sediment was deposited. Caution is, however, required when identifying tillites since deposits produced by mud or debris flows may have similar characteristics. The association of tillites with rock-scoured or striated surfaces is usually considered to be diagnostic of glaciation.

The second indication of a cold or glacial climate is the presence of dropstones in finely bedded (laminated) sediments. Laminated sediments are usually produced by the settling out of sediment from suspension. Periodic variation in the supply of sediment results in the deposition of very fine layers or laminations. Consequently in areas in which laminated sediment is being deposited the only way in which large stones can be deposited without disrupting the laminae is by their vertical deposition from rafts of ice. These rafts may be icebergs or simply sea or lake ice which entrains debris from a shore by freezing. Dropstones alone cannot be used as evidence of glaciation, because of the variety of other possible rafting agents, such as tree roots (Box 10.1).

5.3.4   Reconstructions Based on Climatic Models

Given some basic information about the palaeoclimate of a period, further detail and understanding of its climate can be obtained from computer models. There are two main types of climatic model: (1) simple parametric models, which examine the inter-relationship of a few specific climatic variables; and (2) general circulation models which attempt to simulate the whole of the Earth's atmospheric circulation. Models are used not only to predict climatic patterns and behaviour but also to explore the causes of climatic change.

5.4   Palaeoecology

Palaeoecology is the study of the interaction of ancient organisms with their environment and with each other. Study of the palaeoecology of a group of organisms allows deductions to be made about the nature of their environment.

Fossils form a significant component of the sedimentary record. Today, animals and plants inhabit most environments from the polar wastes to the interstices between grains of sand. The relationship of living organisms to each other and to their environment is known as ecology. It is possible with careful observation to deduce the relationship of the once living and now fossilised organisms within their ancient environment—palaeoecology—by inference from their morphology and by comparison with their living relatives. In essence this is the same process that we applied to the study of facies. In this way by observing the overall characteristics of a body of rock and by comparing it with the known products of contemporary environments we can make deductions about the nature of ancient environments.

Deductions about palaeoenvironment can be made at two levels, from individual fossils and from the interactions of fossils. Study of individuals helps us to understand their life history better, and it is possible to identify in some well-preserved fossils whether an organism has been attacked by a predator, or whether an organism was suffering from a parasitic infestation or illness. A good example of inference about palaeoenvironment from an individual is that of the preservation of insects in amber (fossil tree sap) which not only killed the organism, but also preserved it. Information is therefore available about the nature of the forest environment from both the insect and the amber.

Groups of individuals, and particularly groups of individuals of different species carry the greatest amount of environmental information. The fundamental 'unit' of ecology is the community, that is, a group of two or more species which occupy the same habitat. Such communities are controlled today by a series of interacting, environmentally limiting factors. The relative distribution and diversity of organisms and their interaction with each other is based upon an ability to tolerate the limiting factors present. Examples of such limiting factors include oxygen availability, salinity, substrate, turbulence, water depth, temperature and food supply. It is possible to distinguish today those organisms capable of surviving in a range of environmental conditions (eurytopic organisms), and those who have a narrow environmental tolerance range (stenotopic organisms). As such, in any given assemblage of fossils it is actually the presence of the stenotopic organisms which defines the environment, given the wide tolerance range of the eurytopic ones. Clearly a full understanding of the ecological tolerance of fossil organisms is desirable, and is usually achieved through comparison with nearest living relatives.

Good examples of stenotopic organisms today are the hermatypic, or reef building corals. These only flourish in reefs where the sea water is: (1) of normal salinity; (2) of shallow depth (< 20 metres); and (3) with an optimum temperature of 20°–29°C (Figure 5.19). As these factors are changed the coral community either flourishes or flounders, and clearly the identification of hermatypic fossils in the record is a powerful tool in interpreting palaeoenvironments. Problems arise, however, with particularly ancient organisms without clearly identifiable nearest living relatives, and even where such relatives are present, there is no guarantee of absolute similarity and ecological tolerance. However, although caution is warranted, if the limiting factors can be deduced from the fossils present it is possible to obtain a powerful tool with which to interpret the pattern of changing environments within the stratigraphical record.

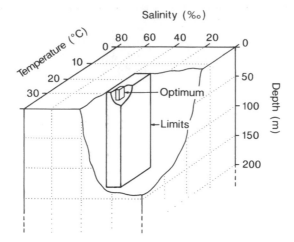

Figure 5.19 *Corals and environmental sensitivity. This diagram illustrates the environmental para-meters which limit the growth of present-day reef-building (hermatypic) corals. The optimum growth of hermatypic corals lies within narrow ranges of salinity, temperature and depth. Using uniformitarianism it is possible to predict that fossil corals lived within the same optimum limits, and therefore provide information on the broad palaeoenvironment. [Modified from: Wells (1957) Geological Society of America Memoir 67, Figure 1, p. 1088]*

There are some important problems associated with the use of fossils in this way. First, it is almost impossible for geologists to be certain whether the assemblage of fossils collected are actually representative of a true community, since most soft-bodied organisms are destroyed before they are fossilised. In addition, fossils from different natural communities may be swept together by the action of currents during deposition. There are some exceptions to this, for example in exceptional circum-stances a whole living community may be preserved. Such exceptional finds are known as Conservation Lagerstätten (fossil bonanzas), but unfortunately these do not occur with sufficient frequency to be of use in the day-to-day interpretation of the stratigraphical record (Box 5.6). Second, with increasing geological antiquity fossil organisms become increasingly different from living organisms and consequently modern analogues cease to be available. For example, the organisms of the Late Proterozoic Ediacaran Biota are almost unique and apparently unrelated to most living organisms and consequently little is known about them.

Despite these problems the fossil record can provide useful information about the nature of palaeoenvironments, at all levels. For example, by inference from modern descendants, cephalopods, echinoderms and corals are only found in marine envir-onments. By contrast, oysters and mussels are at home in intertidal environments, with uncertain salinities. These simple tools are widely used in both field and labor-atory situations, where judgement on the marine or non-marine nature of the facies may be of paramount importance. We can also indentify characteristic biofacies. Fossil assemblages indicative of oxygen-poor bottom waters or those which typically

Box 5.6
Lagerstätten

The fossil record is usually considered to be incomplete. For example, animals with hard parts, such as shells and bones are more likely to survive the rigours of the process of fossilisation than those animals that are wholly soft bodied. For the most part, therefore, fossil assemblages are thought to be rarely representative of the once living communities. Exceptionally, fossils are found in remarkable concentrations (concentration deposits) or with extraordinary preservation of soft parts or of soft-bodied organisms (conservation deposits). Such finds were referred to as Fossil-Lagerstätten by the German palaeontologist Adolf Seilacher, which when translated means 'fossil bonanzas'. Conservation Lagerstätten are exceptionally important as they provide rare glimpses of ancient communities and provide a greater understanding of the biology of fossil organisms. Famous Lagerstätten include the Cambrian Burgess Shale in British Columbia, with its exceptional preservation of soft-bodied organisms, the Devonian Ryhnie Chert of Scotland, which preserves an important early terrestrial biota, and the German Solnhofen Limestone, of Jurassic age, which preserves the evidence of the first bird, *Archaeopteryx*. In all of these, preservation has been quick, and decay halted.

Sources: Clarkson, E.N.K. 1993. *Invertebrate Palaeontology and Evolution*. (Third Edition) Chapman & Hall, London. Whittington, H.B. and Conway Morris, S. (Eds) 1985. Extraordinary fossil biotas: their ecological and evolutionary significance. *Philosophical Transactions of the Royal Society*, Series B **311**, 1–192.

form reefs are examples of biofacies (Box 5.7). We have already seen that fossils aid us in inferring palaeoclimate, but they can also be of use in determining palaeogeography. For example, the Early Palaeozoic brachiopod assemblages of Wales have been interpreted in terms of water depth and consequently the changing assemblage of brachiopods has enabled shorelines for the Early Palaeozoic to be constructed through time, an example which is described more fully in Section 7.2.3.

Amongst the most important fossils for palaeoecological interpretation of the stratigraphical record are trace fossils (Figure 5.20). The study of trace fossils is now commonly called ichnology. Trace fossils represent the tracks, trails and feeding traces of living organisms (including droppings, or coprolites). Although it is extremely rare for the organism to be found in association with its burrow, or feeding trace, it is clear that there is an association of certain types of trace with certain types of environment (Figure 5.20). Trace fossils are classified on the inferred action of their producer: we can recognise resting traces, movement traces, feeding traces, dwelling traces, and so on. Most importantly, they appear to have a correlation with water depth, essentially a function of their facies dependence, and consequently depth related trace fossil biofacies can be recognised (Box 5.8). Trace fossils have a significant advantage over body fossils as they are rarely eroded out of their host rock and reworked into younger deposits, and they are therefore reliable indicators of their environment. They are also conservative, and it has been recognised that broadly similar traces have occupied the same facies throughout the Phanerozoic. For all these reasons trace fossils are extremely valuable tools in the interpretation of the palaeoenvironment of sedimentary sequences, and often, they give a fuller picture of the nature of the biological community than just the shelly fossils alone.

Box 5.7
Oxygen related biofacies

Rhoads and Morse (1971) were first to suggest that the observed sensitivity of organisms to dissolved oxygen could be used to interpret the amount of dissolved oxygen that was present in ancient sedimentary environments. They recognised three oxygen-related biofacies, which they called *anaerobic, dysaerobic* and *aerobic*, based on the levels of dissolved oxygen as illustrated below. In anaerobic facies, bottom-dwelling organisms were absent, consistent with a total lack of oxygen. With dysaerobic facies, some sediment feeding organisms were seen to rework sediments, although bottom-dwelling organisms were largely absent. This is consistent with very low levels of dissolved oxygen, capable of sustaining very little life. Aerobic facies contains a great variety of marine organisms, both within and on top of the sediment, consistent with increased levels of oxygen. Successful interpretation of black shale sequences using the biofacies concept has enabled geologists to give an estimate of the oxygenation of the basin through geological time.

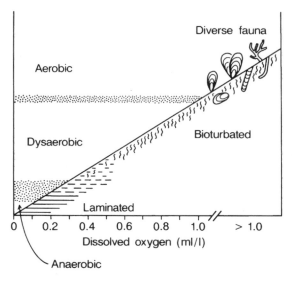

Source: Rhoads, D.C. and Morse, J.W. 1971. Evolutionary and ecologic significance of oxygen-deficient basins. *Lethaia* **4**, 413–428. [Modified from: Rhoads and Morse (1971) *Lethaia* **4**, Figure 5, p. 421.]

5.5 Summary of Key Points

- Sedimentary facies are defined as the sum total of the characteristics of a body of sedimentary rock, which allow the environment of deposition to be deduced. Those characteristics are commonly the geometry of the rock body, its lithology, sedimentary structures and fossil content.
- Facies are not static: the relative position of facies can change in response to sedimentary or autocyclic mechanisms, such as the increased or decreased input of sediment into a sedimentary system. In such systems, the vertical facies variation is equivalent to the lateral facies variation at a specific time (Walther's principle).

Figure 5.20 *Photographs of trace fossils.* **A:** *Skolithos, a simple dwelling trace common in nearshore environments;* **B:** *Diplocraterion, a u-shaped dwelling and feeding burrow common in shelf environments;* **C:** *Thalassinoides, a branching burrow system common in shelf environments; and* **D:** *Palaeodictyon. A complex feeding trace found in deeper water settings [Photographs: P. Doyle]*

- External or allocyclic influences, in particular sea level changes, have a profound influence on the distribution of facies associations. Therefore transgressive (onlapping) and regressive (offlapping) facies associations may be deduced from the stratigraphical record.

- Relative sea level at the continental margin can be deduced from the interpretation of seismic sections. Sedimentary packages can be recognised on these sections and their relative position on or offshore (their relative onlap) can be deduced. This provides a key to relative sea level.

- Sequence stratigraphy recognises that packages of sediment at the continental margin are commonly bounded by unconformities, which are probably genetically related to sea level changes. Given global sea level changes, sequence stratigraphy is a tool for the correlation of large scale sedimentary packages.

- Small scale palaeogeography or the ancient geographies of specific time intervals can be interpreted directly from the association of facies in the stratigraphical record.

- Palaeoclimate may be deduced directly from the study of lithology, palaeontology and geochemical analysis of rocks.

Box 5.8
Bathymetry from trace fossils

It is now commonly accepted that trace fossil assemblages in marine sedimentary rocks are related to the relative depth of accumulation of the sedimentary body. This follows the work of Adolf Seilacher, who recognised that there were a relatively small number of communities of trace fossils which appear limited to different sedimentary facies. Seilacher recognised five distinct assemblages. In coarse, shallow, marine sediments the traces were largely simple dwelling traces of worms (e.g. *Skolithos*). Within shelf marine sediments the traces were indicative of a range of crawling (e.g. *Cruziana*), feeding and more complex dwelling traces. Turbidite sequences commonly contained the complex sediment-mining trace *Zoophycos*, while submarine fan sediments were characterised by complex winding, surface grazing traces (e.g. *Nereites*). Although it is now accepted that the *Skolithos*-type traces are capable of developing in coarse sediments offshore, the *Skolithos*, *Cruziana*, *Zoophycos* and *Nereites* trace fossil biofacies (ichnofacies) are accepted as representative of progressively more offshore environments, as illustrated. These ichnofacies are therefore powerful tools in palaeoenvironmental analysis.

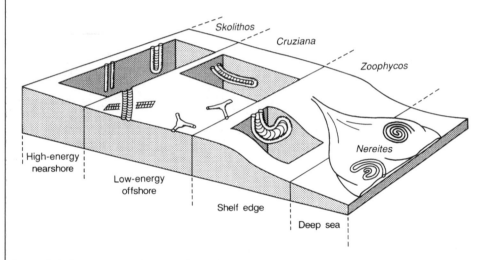

Source: Seilacher, A. 1967. Bathymetry of trace fossils. *Marine Geology* **5**, 413–428. [Modified from: Briggs and Crowther (1990) *Palaeobiology: a Synthesis*, Blackwell, Figure 2, p. 358.]

- Palaeoecology or the ecological relationship of fossils to their ancient environment and to each other can be deduced from the facies relationships of fossils, and is a powerful tool in the reconstruction of ancient environments.

5.6 Suggested Reading

Selley (1985) is a nicely written, easily read and very informative book on the nature of sedimentary facies. Duff (1993) gives a good account of sedimentary environments while Collinson and Thompson (1989) is a very important source on the nature and formation of sedimentary structures. Good general accounts are given by Boggs

(1987) and Prothero (1990). Hallam (1992) provides in-depth coverage on sea level changes and provides a useful critique of sequence stratigraphy (see also Hallam, 1998). Emery and Myers (1996) and Vincent et al (1998) provide a modern review of sequence stratigraphy. Tucker (1988) is a very good handbook providing information on most of the topics in this chapter. Ager (1963) is quite an old text but still important as an introduction to palaeoecological principles. Further information on many of the concepts presented in this chapter is given in Brown et al (1992). Principles of facies analysis, palaeoclimatology and palaeoecology are reviewed by Pirrie (1998), Francis (1998) and Doyle and Bennett (1998). Papers in Brenchley (1984) provide a useful overview of the role of fossils in determining ancient climates.

References

Ager, D.V. 1963. *Principles of Palaeoecology*. McGraw-Hill, New York.

Boggs, S. 1987. *Principles of Sedimentology and Stratigraphy*. Merrill, New York.

Brenchley. P.J. (Ed.) 1984. *Fossils and Climate*. John Wiley, Chichester.

Brown, G.C., Hawkesworth, C.J. and Wilson, R.C.L. (Eds) 1992. *Understanding the Earth*. (Second Edition). Cambridge University Press.

Collinson, J.D. and Thompson, D.B. 1989. *Sedimentary structures*, (Second Edition). Unwin Hyman, London.

Doyle, P. and Bennett, M.R. 1998. Interpreting palaeoenvironments from fossils. In Doyle, P. and Bennett, M.R. (Eds) *Unlocking the Stratigraphical Record*. John Wiley, Chichester, 441–470.

Duff, P. McL. D. (Ed.) 1993. *Holmes' Physical Geology*. Chapman & Hall, London.

Emery, D. and Myers, K. (Eds). 1996. *Sequence Stratigraphy*. Blackwells, Oxford.

Francis, J.E. 1998. Interpreting palaeoclimates. In Doyle, P. and Bennett, M.R. (Eds) *Unlocking the Stratigraphical Record*. John Wiley, Chichester, 471–490.

Hallam, A. 1992. *Phanerozoic sea-level changes*. Columbia University Press, New York.

Hallam, A. 1998. Interpreting sea level. In Doyle, P. and Bennett, M.R. (Eds) *Unlocking the Stratigraphical Record*. John Wiley, Chichester, 421–440.

Pirrie, D. 1998. Interpreting the record: facies analysis. In Doyle, P. and Bennett, M.R. (Eds) *Unlocking the Stratigraphical Record*. John Wiley, Chichester, 395–420.

Prothero, D.R. 1990. *Interpreting the Stratigraphic Record*. Freeman, New York.

Selley, R.C. 1985. *Ancient Sedimentary Environments*, (Third Edition). Chapman & Hall, London.

Tucker, M.E. 1988. *The Field Description of Sedimentary Rocks*. Open University Press, Milton Keynes.

Vincent, S.J., Macdonald, D.I.M., and Gutteridge, P. 1998. Sequence stratigraphy. In Doyle, P. and Bennett, M.R. (Eds) *Unlocking the Stratigraphical Record*. John Wiley, Chichester, 299–350.

6

The Evolution and Closure of Sedimentary Basins: The Role of Plate Tectonics

In the preceding chapters we have introduced the concept of a basic stratigraphical tool kit with which it is possible to determine the history of rock units, and interpret each unit in terms of ancient Earth environments. This chapter now considers the distribution of rocks across the surface of the Earth in the context of the development of the Earth's major structural components and their interaction.

6.1 The Mystery of Mountain Building

Early geologists thought in terms of the interpretation of local and regional rock successions, and their world view tended to be based upon an extrapolation of these local geologies. However, throughout the nineteenth century there was an increasing emphasis on the comparison of the rock successions over very large areas as ideas of a global pattern started to emerge. This shift in concepts was possible with the evolution and increasing sophistication of the stratigraphical tool kit.

This large scale or global perspective led to the discovery that many of the thickest sequences of sedimentary rocks tended to be concentrated in long, linear belts. These belts occur along the margins of ancient continental crust known as cratons and have been heavily folded and faulted. These belts of deformed rock form the Earth's main tectonic belts (Figure 6.1). They consist of rocks which were deposited in deep

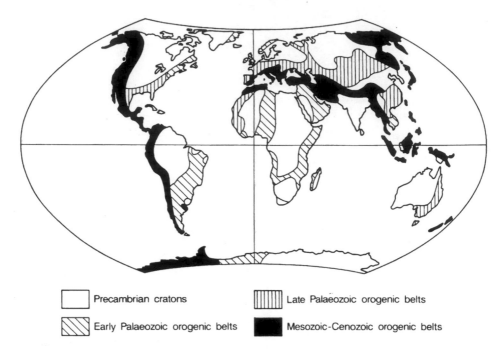

Figure 6.1   *The world's major tectonic belts. [Modified from: Anderton et al (1979) A Dynamic Stratigraphy of the British Isles, Allen & Unwin, Figure 1.1, p. 3]*

sedimentary basins and which have been subsequently folded, faulted, thrust, intruded by igneous rocks and metamorphosed. The uplift of these belts of deformed sediments has produced long chains of mountains which have been carved and shaped to various degrees by erosional processes. The formation of these sedimentary basins and their subsequent compression provided one of the most enduring mysteries for the early geologist. Ever since James Hutton first discovered the presence of cycles of uplift, erosion and deposition within the stratigraphical record, the quest for the mechanism of uplift or mountain building was on.

In the middle of the nineteenth century geologists began to interpret these tectonic belts in terms of geosynclines. Geosynclines were seen as regional troughs or basins which gradually subside as they are filled by sediment. This subsidence led to the intrusion of igneous rocks in the basin. Folding and metamorphism took place when a thick pile of sediments and igneous rocks had accumulated, and this deformation eventually led to the formation of a mountain chain. The mechanism by which deformation was produced was never adequately explained. In the following hundred years the essential elements of the geosynclinal theory were modified and refined. However, it was not until the development of the theory of plate tectonics in the mid-1960s that a coherent picture of the formation of sedimentary basins and their deformation into tectonic belts was obtained. Plate tectonics has unlocked the mystery of mountain building, and continues to influence our understanding of the geological evolution of the Earth.

6.2 Continents Adrift

The idea that the continents may have moved across the Earth's surface through geological time was explored by several geologists in the nineteenth and early twentieth centuries. The most convincing evidence was the remarkable fit between the Atlantic coasts of Africa and South America. However, the hypothesis of continental drift was largely developed by one man, the German meteorologist Alfred Wegener. Wegener suggested in the early twentieth century that the Earth's continents had at one time been joined in two supercontinents. Wegener used as his evidence the fit of the continents already alluded to, but he also used many elements of the stratigraphical tool kit. In particular he took two lines of essentially stratigraphical evidence: fossil distribution, and lithological similarity.

Wegener studied the distribution of fossil land plants and animals to aid in his interpretations. In particular, he studied the distribution of the plant *Glossopteris*, the leaf remains of which were relatively common in the Permian successions of the Southern Hemisphere continents. Wegener reasoned that in order for *Glossopteris* leaves to be found in the widely spaced continents of the Southern Hemisphere the continents must once have been joined. Using this evidence Wegener grouped all of the southern continents, together with India into the supercontinent of Gondwanaland, now more commonly shortened to Gondwana. Wegener also studied the distribution of major geological bodies, such as crystalline basement complexes and mineral deposits. Using these lithological criteria Wegener was able to demonstrate that the fit predicted by map estimates was confirmed by the alignment of geological complexes on either side of the Atlantic Ocean. Wegener presented his findings in his book *Die Entstehung der Kontinente und Ozeane* (The origin of continents and oceans) published in 1915. Although some of Wegener's contemporaries accepted his ideas, these pieces of largely stratigraphical evidence were not considered powerful enough to convince sceptics, who felt that the underlying cause of continental movements was not clearly explained, let alone possible.

The theory awaited the discovery of palaeomagnetism and the development of oceanography before its wider acceptance by the scientific community. As discussed in Section 5.2, palaeomagnetism employs the principle that in molten igneous rocks, or unlithified sediments, any magnetic particles will align themselves with the Earth's magnetic field. This magnetic record is stored within igneous rocks when they cool and within sediments when they become lithified. Deviation in the alignment of such palaeomagnetic particles from the current direction of the Earth's magnetic field shows that the continents have moved. In the 1960s two Cambridge scientists, Fred Vine and Drummond Matthews, discovered that on either side of the mid-Atlantic ridge there was a series of linear magnetic anomalies. In fact it was observed that strips of ocean crust had alternating magnetic orientations. These observations were explained by Harry Hess in terms of a sea floor spreading model by which new oceanic crust forms along mid-ocean ridges as the two halves of an ocean move apart. From these simple observations the theory of plate tectonics developed and is now seen by most as the ultimate control on the stratigraphical record. For this reason the principles of plate tectonics and its role in developing sedimentary basins are examined in this chapter.

6.3   Plate Tectonics

The plate tectonic model proposes that the surface of the Earth consists of a series of relatively thin, but rigid plates which are in constant motion (Figure 6.2). The surface layer of each plate is composed of either oceanic crust, continental crust or a combination of both, while the lower part consists of the rigid upper layer of the Earth's mantle (lithospheric mantle). The rigid plates pass gradually downwards into the plastic layer of the mantle, known as the asthenosphere. The plates vary in area from 10^6 to 10^8 km^2 and may be up to 70 km thick if composed of oceanic crust or 150 km if incorporating continental crust (Figure 6.2). There are seven major plates (10^8 km^2), eight intermediate ones (10^6–10^7 km^2) and over 20 small ones ($< 10^6$ km^2). Plates can move at up to 100 mm per year, although the average rate of movement is about 70 mm per year. Much, but not all, of the Earth's tectonic, volcanic and seismic activity occurs at the boundaries of neighbouring plates. In practice, plate boundaries are recognised by the distribution of volcanoes and earthquakes, which dramatically emphasise the internal activity of the Earth. There are three types of plate boundaries: (1) divergent boundaries; (2) convergent boundaries; (3) transform boundaries.

1. Divergent plate margins.   At this type of boundary plates move apart and new oceanic crust is formed in the gap between the two diverging plates (Figure 6.3). Plate area is therefore increased at divergent boundaries. Plate movement takes place laterally away from the plate boundary, which is usually marked by a ridge or rise. This ridge or rise may be offset by transform faults (Figure 6.3). Today, most divergent margins occur along the central zone of the world's major ocean basins and the process by which the plates move apart is referred to as sea-floor

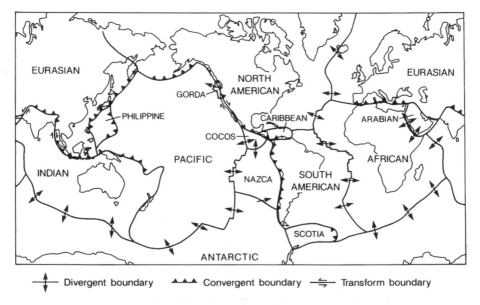

Figure 6.2   *Map of the major lithospheric plates. The various types of plate boundary are also shown*

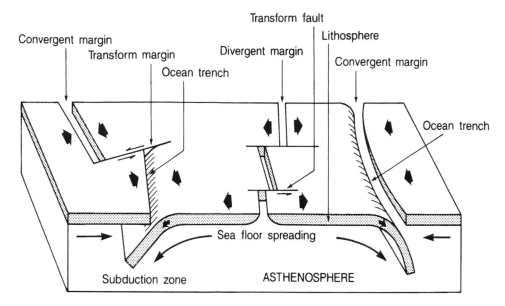

Figure 6.3   *Major plate boundaries. At divergent plate margins plates move apart by sea floor spreading. At convergent plate margins one plate descends beneath the other along a subduction zone and is consumed by the lower mantle. At transform margins plates simply slide past one another. The Earth's lithospheric plates rest on a more mobile layer known as the asthenosphere. [Modified from: Isacks et al (1968)* Journal of Geophysical Research **73**, *Figure 1, p. 5857]*

Box 6.1
Plate tectonics: the final proof

In 1984 scientists at NASA released the first direct measurement of plate movements on the surface of the Earth. By measuring the separation of points on the Earth's surfaces over a period of time it is possible to determine the movement of plates to an accuracy of a few centimetres. These measurements are made with astronomical telescopes and there are about 20 stations around the world which are equipped to make this type of measurement. The measurements are made in one of two ways. In the first method astronomers measure the time taken for two radio telescopes, on different plates, to receive the same signal from outer space. From this it is possible to estimate the rate at which the two telescopes move apart if regular measurements are made. Astronomers have been making measurements of this sort across the Atlantic for over 10 years and have estimated that it widens by about 150 mm every year. The second method involves bouncing laser signals off a satellite equipped with a reflector. The distance of the laser from a satellite is determined by measuring the time taken for the laser beam to be reflected from a satellite and returned to its source. Repeated measurement over a number of years allows the rate at which the plate beneath the laser moves to be measured in relation to the position of the satellite. For example, laser stations have monitored the San Andreas fault for over 11 years and have found that the two plates which form this transform margin move past each other at rate of 60 mm every year.

Source: Henbest, N. 1984. Continental drift: the final proof. *New Scientist* **105**, 6.

spreading. The Mid-Atlantic Ridge and East Pacific Rise provide good examples of this type of plate margin. The rate at which each plate moves apart from a divergent margin varies from less than 50 mm per year to over 90 mm per year and can be determined from the pattern of magnetic anomalies either side of a spreading ridge. Either side of a spreading centre, weak magnetic anomalies 5–50 km wide and hundreds of kilometres long can be identified. As molten rock cools between diverging plates the magnetic minerals present align themselves with the orientation of the Earth's magnetic field at that time. The polarity of the Earth has changed at regular intervals throughout geological time: magnetic north has alternated between the Arctic (normal polarity) and the Antarctic (reversed polarity). Consequently, sections of crust formed during a period of normal polarity have a palaeomagnetic remnance which is oriented towards today's magnetic north, while a section of crust formed during a period of reversed polarity does not. These long linear strips of magnetic anomalies form a symmetrical pattern either side of a spreading centre, and it was this pattern that was first interpreted by Vine and Matthews. In fact, a record of the changes in the Earth's magnetic polarity has been established and dated for the Cenozoic and is the basis for magnetostratigraphy, a method of sequencing rocks relative to the record of magnetic reversals. This stratigraphical record, in conjunction with the magnetic stripes found either side of a spreading ridge allow the rate and pattern of sea floor spreading to be examined.

2. Convergent plate boundaries.   At a convergent boundary two plates are in relative motion towards each other (Figure 6.3). One of the two plates slides down below the other at an angle of around 45° and is incorporated into the Earth's mantle along a subduction zone. The path of this descending plate can be determined from an analysis of deep foci earthquakes and the initial point of descent is marked on the surface by a deep ocean trench. Plate area is reduced along the subduction zone. The detailed morphology of convergent margins depends upon whether the converging plates are both composed of oceanic crust or whether one or both of the plates consists of continental crust (Figure 6.4).

When two plates of oceanic crust collide a volcanic island arc may form (Figure 6.4). As one of the plates is subducted beneath the other it begins to melt at a depth of between 90 and 150 km and the resulting magma rises to the surface above the subduction zone to form a chain or arc of volcanoes. The edge of the plate which is not descending is therefore marked by a chain of volcanic islands. If one traversed this type of plate boundary, one would first cross a deep ocean trench and then a volcanic island chain. The distance between the volcanic island and the trench is determined by the angle at which subduction occurs. If a plate descends steeply ($>45°$) then the distance from the trench to the volcanic arc will be small, but if subduction is shallow ($<45°$) there will be a broader gap between the trench and arc.

Where a plate composed of oceanic crust meets one of continental crust the former is subducted (Figure 6.4). The molten rock produced in the subduction zone rises to form either a volcanic arc or alternatively, if the edge of the continental plate is compressed, a chain of volcanoes and mountains are produced along the edge of the continental plate.

A: Ocean plate–ocean plate convergence   B: Ocean plate–continental plate convergence

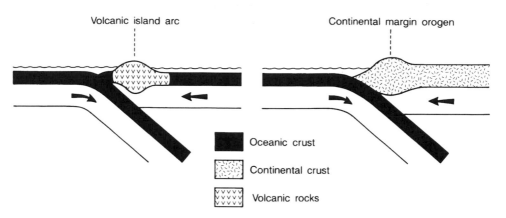

Figure 6.4   *The two main types of convergent plate margin.* **A:** *This situation occurs where an ocean plate converges with another ocean plate and a volcanic island arc is created.* **B:** *This situation occurs where an ocean plate converges with a continental plate. Here the ocean plate is subducted beneath the continental margin which may be compressed to form a mountain belt.*

3. Conservative or transform margin.   At a conservative or transform margin two plates move laterally past each other and oceanic crust is neither created, nor destroyed. The most famous example of this type of boundary is the San Andreas fault system in southern California.

There are two fundamental assumptions which underlie the plate tectonic model. First it assumes that the surface area of the Earth has not changed significantly with respect to the rate at which new oceanic crust is generated at divergent margins. Second, that there is little internal deformation within plates relative to the motion between them. If the Earth's surface has increased significantly then sea-floor spreading at divergent margins could occur without the destruction of plates at subduction zones, however, most evidence suggests that the Earth's surface area has not changed significantly during the last 500 Ma.

In spite of a general acceptance of the plate tectonic model, the question of what causes plates to move has yet to be fully resolved. Four main hypotheses have been put forward to explain this problem. The first hypothesis suggests that flow in the mantle is induced by convection currents which drag and move the lithospheric plates above the asthenosphere. Convection currents rise and spread below divergent plate boundaries and converge and descend along convergent margins. The convection currents result from three sources of heat: (1) cooling of the Earth's core; (2) radioactivity within the mantle and crust; and (3) cooling of the mantle. The convection hypothesis has been proposed in several different forms throughout the last 60 years, but in each case the size of the convection currents and the amount of mantle that is in motion has been different. Recent work has questioned the convection hypothesis.

There is growing evidence that mantle convection does occur at a range of different scales, but recent work suggests that the pattern of divergent plate boundaries – points of convective upwelling according to the convection hypothesis – on the Earth's surface is not consistent with the pattern of heat convection within the Earth's upper mantle. Moreover, the presence of transform faults and geophysical properties of divergent boundaries are also inconsistent with convection. The second hypothesis invokes the injection of magma at a spreading centre pushing plates apart and thereby causing plate movement. The third possible mechanism involves gravity. Here it is suggested that oceanic lithosphere thickens as it moves away from a spreading centre and cools, a configuration which might tend to induce plates to slide, under

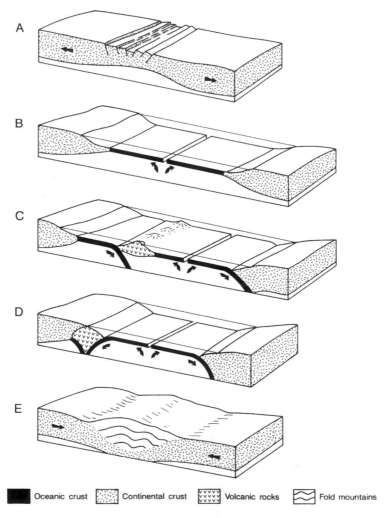

| ■ Oceanic crust | ░ Continental crust | ⩗ Volcanic rocks | ∿ Fold mountains |

Figure 6.5   *The Wilson Cycle. This cycle shows the evolution and closure of an ocean basin to produce a linear chain of fold mountains. **A:** Rifting of continental crust. **B:** Formation of new ocean crust by sea floor spreading. **C & D:** Closure of ocean by subduction. **E:** Following the subduction of all the oceanic crust the two continental plates collide to produce an orogenic belt*

the force of gravity, from a divergent margin towards a convergent one. The final hypothesis suggests that a cold dense plate descending into the mantle at a subduction zone may pull the rest of the plate with it and thereby cause plate motion. This last mechanism has gained support in recent years, but there is still widespread disagreement about the precise mechanism involved.

The plate tectonic model provides a framework with which to understand the evolution, growth and closure of ocean basins. Ocean basins are formed by the rifting and break-up of continents and grow through the process of sea-floor spreading. Within the life time of an ocean, subduction will begin and may come to dominate the tectonic processes. As a consequence the ocean will shrink as more sea floor is lost through subduction than is created by sea-floor spreading. Eventually the ocean will close and the continents along its margins will collide and fold the sediments which accumulated in the former ocean to form a chain of mountains. The line of closure is known as the suture. This complete cycle in the life of an ocean is known as the Wilson Cycle and today's oceans can be seen as part of this cycle (Figure 6.5). It is important, however, to note that the Wilson Cycle is essentially a descriptive concept as a given ocean basin may not necessarily follow the full cycle.

In summary, the plate tectonic model provides a mechanism by which: (1) continents can move across the surface of the globe; (2) patterns of volcanism can change and shift across the globe as plates and their boundaries evolve and move; (3) new oceans may grow and different sedimentary basins evolve; and (4) oceans and sedimentary basins close and are deformed to produce mountains. These are mechanisms which are essential to understanding the Earth's stratigraphical record.

6.4 Plume Tectonics

Over the last 20 years a growing body of evidence has emerged to suggest that plate tectonic activity may not operate with a uniformity of rate, but may be subject to periodic accelerations of activity, a phenomenon which has, in some quarters, has been referred to as pulsation tectonics. In the last 10 years these episodes of accelerated activity have increasingly been explained by plume tectonics. Plume tectonics involves the upwelling of large bodies of molten rock within the mantle and their interaction with the lithosphere. These superplumes appear to originate at the junction between the core and mantle at a depth of over 2900 km. They are hundreds of kilometres in diameter and may be 250° to 300°C hotter than surrounding mantle. The origin of these plumes is uncertain but they appear to be generated by instability at the core- mantle boundary. Each superplume rises through the mantle before encountering the lithosphere, where it mushrooms out below the lithosphere causing it to dome, rift and melt. This process may be associated with the production of large amounts of flood basalt, accelerated rifting and subsequent plate divergence.

Early models of these superplumes suggested that like today's hot spots in plate interiors (small upper mantle plumes), superplumes were simply superimposed and independent of plate tectonics. What is now emerging is the idea of a 'whole Earth' tectonic model in which plate tectonics both influences, and in turn is influenced by, plume tectonics. The details of such models are still unclear, but a recurrent theme is

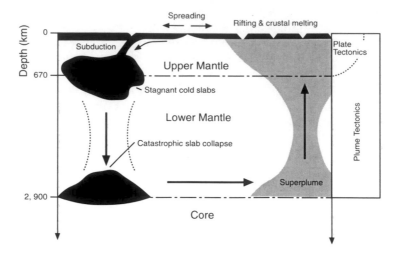

Figure 6.6   *The idea of plume tectonics. Plate tectonics provides via subduction cold slabs of oceanic crust which stagnate between the Upper and Lower Mantle, before collapsing towards the outer core. In doing so they may initiate a thermal plume. [Modified from: Huggart (1997) Environmental Change, Routledge, Figure 3.8, p. 74]*

the idea that subduction of cold slabs of ocean crust may generate superplumes. Slabs of relatively cold oceanic crust appear to accumulate below subduction zones in the upper mantle at a depth of 600 km (Figure 6.6). Many people now suggest that plate tectonics is driven by this subduction pulling plates across the Earth. As this cold ocean crust accumulates in the upper mantle it becomes subject to catastrophic collapse—known as avalanching—into the lower mantle, towards the mantle–core boundary. A consequence of this collapse is the rise of a superplume from the mantle–core boundary. In this way material is recycled through the mantle, and heat is convected from the core to the surface (Figure 6.6). If one imagines the Earth's core to be a large heater which must release a given amount of heat, then instead of releasing this heat in small amounts over a prolonged period, it does so in short bursts of great magnitude, via superplumes.

In the type of geotectonic model just outlined there is clearly an interplay between plate tectonics and plume tectonics, but it must be emphasised that the finer details of this 'whole Earth' tectonic model have yet to be resolved. What is clear, however, is that plate tectonics operates on the Earth's surface, and that it is periodically accelerated and influenced by mantle plumes. The new concepts are simply refining our understanding of the large scale mechanism of plate tectonics.

6.5   The Evolution of Sedimentary Basins

Most sedimentary rocks are deposited in some form of basin, the size and shape of which is controlled by the processes of plate tectonics. As such the preservation of the sedimentary record is primarily a function of plate tectonics. Present day sedimentary basins provide analogues with which to interpret the depositional environments of sedimentary rocks in the stratigraphical record and it is therefore important to interpret the record from this perspective. The processes which operate at plate margins produce a wide range of different types of sedimentary basins, but in general five broad categories can be recognised (Table 6.1) each of which may be associated with a distinct type of sedimentary content (Table 6.2).

1. Basins and divergent plate margins.   Two types of basin occur in association with plate divergence and therefore with the first half of the Wilson Cycle (Figure 6.5). The formation of a rift valley is the first stage in the separation of a piece of continental crust into two plates (Figure 6.7). As the two pieces of continental crust move apart the rift-valley increases in width and may subside toward sea

Table 6.1   *Types of sedimentary basin*

1. Basins produced by extension at or close to divergent plate margins
   - Terrestrial rift valleys within continental crust
   - Passive continental margins in mid-plate locations but at the continent–ocean interface
2. Basins formed at convergent plate margins
   - Trenches formed by subduction of oceanic crust
   - Fore-arc basins which develop between ocean trenches and margins of a continental or island arc
   - Back-arc basins which form behind island arcs
   - Foreland basins formed at continental margins during crustal collision
3. Basins formed along continental transform margins
   - Basins formed along strike-slip fault systems
4. Continental basins unrelated to plate margins
5. Ocean basins

Table 6.2   *Typical sediments within different sedimentary basins. [Based on information in: Dott & Batten (1988)* Evolution of the Earth. *McGraw-Hill; Nichols (1993). In: Duff (Ed.) Holmes' Principles of Physical Geology. Chapman & Hall]*

Basin Type	Characteristic Sediment	Depositional Environment
Trench	Fine sediment overlying ocean-floor basalts	Deep-marine
Fore-arc	Heterogeneous gravels, sands and muds derived from erosion of volcanic, metamorphic and granitic rocks of the adjacent volcanic arc or continental margin	Non-marine to marine
Foreland	Heterogeneous gravels, sands and muds derived from the orogenic belt	Mostly river and deltaic
Passive margin	Quartz-rich sands and limestones passing seaward to muds	Shallow-marine shelf to deeper-marine
Rift valley	Earliest rock volcanic, overlain by thick gravels and sand; younger rocks may include evaporites and limestones.	Rivers and lakes changing to shallow-marine

A　Symmetrical rift valley

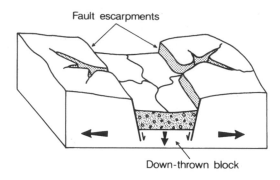

B　Asymmetrical rift valley

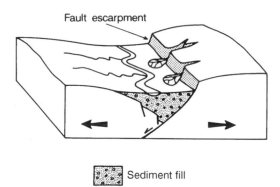

Figure 6.7　*The structure and morphology of rift valleys.* **A:** *A symmetrical rift valley, or graben.* **B:** *An asymmetrical rift valley, or half-graben. [Modified from: Summerfield (1991) Global Geomorphology, Longman, Figure 4.9, p. 92]*

level where it may flood to form a shallow sea. Continued plate divergence will widen this seaway and its floor will gradually become composed of oceanic crust. A sea floor spreading ridge will develop along the central axis of the sea and continued divergence will lead to the evolution of an ocean basin (Figure 6.5). The Red Sea is presently at the first phase, produced by the divergence of the Arabian and African plates, while the Atlantic Ocean represents a later stage in the process, since the North American and European plates have been moving apart for much longer. It is important to note that the formation of a rift valley does not automatically lead to the rupturing of a plate and the formation of a new ocean basin: the process of divergence may stop if the driving mechanism is shut off.

The trailing edges of the continents that result from the formation of a new ocean are sites of little seismic activity and are not associated with plate movement (Figure 6.5). They are therefore referred to as passive margins: the ocean–continent interface is not marked by a subduction zone and both continent and ocean are part of the same plate. Passive margins are normally associated with large continental shelves and shallow coastal waters, which are ideal for accumulation of a large sediment pile. Such sediments tend to spread from these shelf environments into the deeper ocean basins via large sub-aqueous flows known as turbidity currents. The Atlantic seaboards of North America and Europe provide good examples of passive margins.

2. Basins and convergent plate margins.   Four types of basin are associated with convergent plate margins, and are generally related to the second half of the Wilson Cycle (Figure 6.5). The destruction of oceanic crust occurs along subduction zones, which are marked on the surface of the ocean floor by a deep trench (Figures 6.3 and 6.8). Ocean trenches are typically about 2 km deeper than the level of the ocean floor and are often more than 6 km below sea level. Trenches form the deepest parts of ocean basins, for example the Marianas Trench in the Pacific Ocean is more than 10 km below sea level. Ocean trenches act as large traps for sediment produced either on volcanic island arcs or from continental margins. As a plate is subducted, trench sediment may either be taken down into the mantle or alternatively it may be scraped off by the overriding plate and added to it in a series of slices. The repeated addition of slices at the edge of the overriding plate may build a triangular body of sediment along the outer edge of a trench know as an accretionary prism (Figures 6.8 and 6.9). This triangular body of sediment may define a basin between the trench and the continental margin or island arc, known as a fore-arc basin (Figure 6.8). The width of a fore-arc basin is controlled by the distance between the accretionary prism at the trench and the continental margin or island arc, which is determined by the angle of subduction. If the descending plate plunges steeply into the subduction zone

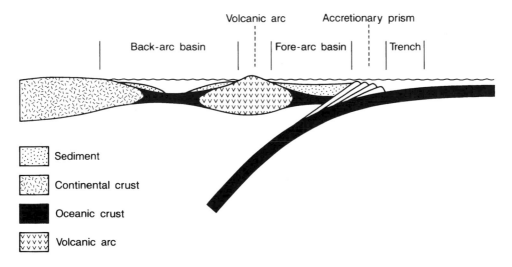

Figure 6.8   *Types of sedimentary basin associated with a subduction zone. Not all these basins may be present along any one margin*

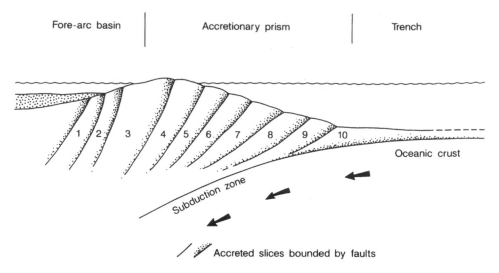

Figure 6.9   *A model of an accretionary prism. As the ocean crust descends slices of trench sediment are accreted to the opposite side of the trench. Each slice is usually fault bounded and the numbers refer to the age of each slice, 1 is the oldest and 10 is the youngest. [Modified from: Nichols (1993). In: Duff (Ed.) Holmes' Principles of Physical Geology, Chapman & Hall, Figure 30.14, p. 710]*

then the fore-arc basin will be narrow and conversely if the angle of subduction is shallow then the fore-arc basin will be wider.

A back-arc basin may form to the rear or landward side of an island arc (Figure 6.8). The geometry of back-arc basins can be quite complex and their size usually depends upon the position of the island arc in relation to other arcs or to the continental margin.

Subduction of crust in a trench can eventually consume an ocean. As the ocean closes it brings two continents together to the point where they collide in the final stage of the Wilson Cycle (Figure 6.5). The collision of these two continents causes intense deformation of the sediments deposited in the former ocean. The deformed sediments form large thrusts and huge folds of rock called nappes which move both upwards and laterally away from the point of collision. These thrust sheets load the crust adjacent to the area of collision causing it to subside under their weight. This loading may form a peripheral basin known as a foreland basin which quickly fills with sediment as the thrust sheets and nappes forming a new range of mountains erode.

3. Basins formed at transform margins.   At transform plate margins crust is neither created nor destroyed. However, movement between the two plates is not smooth and if the plate edges are irregular, zones of local compression (transpression) and extension (transtension) are created as the plates move past one another. Basins formed by transtension are characteristically deep in relation to their size and their margins are associated with steep faults (Figure 6.10). Examples of this type of basin include the Dead Sea rift and the Gulf of California. It is possible to identify a cyclic evolutionary history for this type of basin similar to that recognised for ocean basins by Wilson. The first phase of the cycle involves trans-

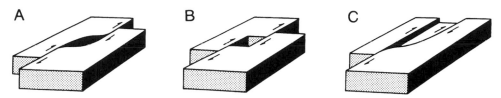

Figure 6.10 *Basins formed along transform margins or strike-slip faults.* **A:** *A 'releasing bend' in a single fault creates a gap when the fault moves.* **B:** *An offset in a fault produces a 'pull-apart' basin.* **C:** *A branch in a strike-slip fault produces a region of extension between two faults. [Modified from: Reading (1980). In: Ballance & Reading (Eds)* Sedimentation in Oblique-slip Mobile Zones, *Special Publication of the International Association of Sedimentologists No. 4, Figure 3, p. 12]*

tension and the formation of a basin which quickly fills with both coarse alluvial gravel and fine-grained marine or lacustrine sediments. Continued extension and basin subsidence leads to the introduction of volcanic rocks in the floor of the basin. In the second phase of the cycle, transpression causes the closure and deformation of the basin.

4. Continental basins unrelated to plate margins. A variety of different types of sedimentary basin may form within a plate composed of continental crust. Of these basins the intracratonic basin is perhaps the most noteworthy. This type of basin is a broad regional downwarp in continental crust, well removed from the tectonic influence of a plate margin. They do not usually involve faulting, although some, but not all of the basins may be centred on local rift valleys. They are commonly oval in shape and are enclosed, with a central point much lower than the surrounding area. The processes by which these broad downwarps form are not entirely clear, although it has been suggested that they result from the activity of thermal plumes in the underlying mantle. Most volcanic activity is confined to plate boundaries, but local hot spots or thermal plumes do occur within the interior of plates. If a thermal plume rises beneath a continent it will cause the overlying crust to thin and stretch to produce a rift valley. If this process continues the crust will separate to form two new plates and a new ocean. However, if the thermal event comes to an end before crustal separation occurs, then a broad crustal down-sagging may result when the heat is removed, which will give rise to an intracratonic basin. Alternatively thermal plumes may elevate, but not cause rifting. At the end of the thermal episode the crust will subside back to its original position. If it has, however, been thinned by erosion while elevated the crust will subside below its original level to form a basin. In this way regional downwarps may form which are not centred on local rift valleys.

5. Ocean Basins. Ocean basins are the largest type of sedimentary basin, although the rate of sedimentation, and therefore the relative volume of sediments, within them is often very low due to the distance of the central parts of the basin from land and a major source of sediment. In most cases sedimentation in these basins takes the form of the continuous rain of pelagic organisms at basin centre, creating fine sedimentary 'oozes' of calcareous and siliceous micro-organisms. Sedimentary volume increases dramatically as the shelf edge and slope are encountered, with fans of clastic sediments from submarine canyons, often fed by density flows.

In conclusion, it is important to emphasise that plate tectonics is a dynamic process and that sedimentary basins may evolve from one type or setting to another through time. For example, as the Wilson Cycle develops a rift valley evolves into an ocean basin. The margins of the ocean basin may then become back-arc or fore-arc basins as subduction of ocean floor commences once it is mature. In referring to a sedimentary basin we are therefore not considering a constant feature but one which changes in character through time.

6.6   The Closure of Sedimentary Basins: the Processes of Mountain Building

Mountain building (orogenesis) involves the compression and deformation of a sedimentary basin. There are four types of mountain building processes: (1) continental margin orogenesis; (2) continental collision; (3) collage collision; and (4) transpression.

1. Continental margin orogenesis.   Not all mountain chains form by the closure of ocean basins and orogenesis may occur along plate margins which are not under-going collision (Figure 6.4). This type of mountain building is referred to as a continental margin orogenesis and involves the production of a mountain chain along a convergent margin where an oceanic plate is descending beneath a plate composed of continental crust. Compression and deformation will occur at the plate margin, if the rate at which plate convergence is occurring exceeds the rate of subduction. The subduction rate is controlled by the rate and angle at which the plate of ocean crust descends beneath the plate of continental crust. This is controlled by the buoyancy of the ocean plate or its temperature – a cold plate, for example, will descend easily. This in turn is determined by the distance between the centre of sea-floor spreading and the subduction zone: the greater the distance the greater the opportunity for the ocean crust to have cooled. If com-pression occurs it leads to buckling and deformation of the leading edge of the continental plate which is deformed into a chain of mountains parallel to the sub-duction zone. The Andes provide an excellent example of this type of orogen. Mountain building here is caused by the buckling of the front edge of the conti-nental plate and by the intrusion of large volumes of molten igneous rock which rises from the subduction zone.

2. Continental collisions.   This involves the production of a range of mountains by the closure of an ocean basin and the collision of two continents along its margin (Figure 6.11). It is this type of orogenic event which is represented by the Wilson Cycle (Figure 6.5). The Himalayas provide the best contemporary example

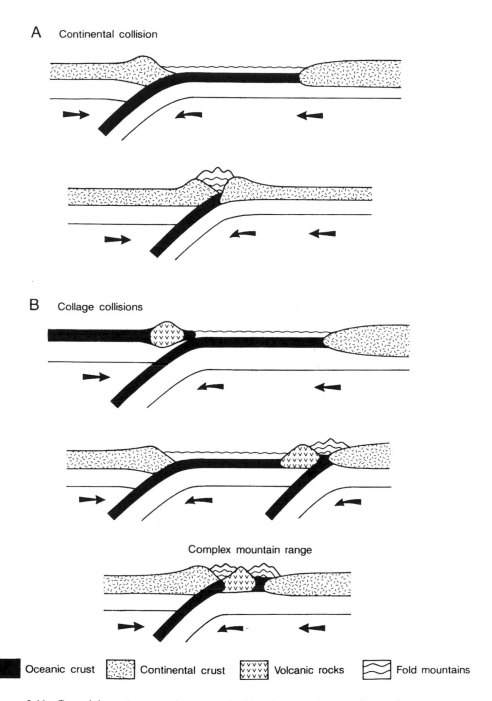

Figure 6.11   *Two of the main types of mountain building (orogenesis).* **A:** *Collision of two continents.* **B:** *An example of a collage collision involving two continents and an island arc*

of this sort of orogenic episode. Over the last 80 million years or so the Indian sub-continent has moved progressively northwards as part of the Indian plate. The ocean between the sub-continent and the Eurasian landmass has slowly closed leading to collision between the Indian sub-continent and Eurasia. The Himalayas were produced by this collision.

3. Collage collisions. Mountain building may occur as a result of the collision of several different plate elements or fragments of ocean floor. At a simple level, island arcs may collide with continents, which in turn may collide with other continents, and in this way build up a chain of mountains in successive stages by the addition of successive distinct elements (Figure 6.11). It is probable that any structure which rises above the general level of the ocean floor, such as seamounts, oceanic islands, basaltic plateaus, island arcs or continental fragments, will be accreted to a continental margin when it reaches a subduction zone, rather than being subducted. In this way a variety of different structures may become jammed together to form part of a mountain belt.

Small slivers of continental crust, known as micro-continents, are commonly involved with this type of collision. Each island arc or micro-continent that is added to the collision is called a terrane and each terrane may have its own stratigraphical record and different fossil assemblages (Box 6.2). Terranes may not only consist of micro-continents and island arcs but also a wide variety of other types of topographic fragments, such as sea mounts, ocean ridges, or accretionary prisms. This type of mountain building is known as collage tectonics because of the wide variety of different elements (the terranes) which are involved. The acceptance of collage

---

Box 6.2
Terranes

A terrane is a mappable structural entity which has a stratigraphical sequence and an igneous, metamorphic and structural history quite distinct from those of adjacent units. Terranes are separated from each other by a distinct structural break, generally a fault. They are commonly referred to as allochthonous which means that the association of rocks that comprise each terrane were not formed in its present position, and that it has been moved. Many modern and ancient mountain chains appear to consist of a sequence of terranes which have been accreted to a continental margin at a subduction zone to form part of either a continental orogen or the core of an orogenic belt produced by the collision of two continents.

The Western Cordillera of North America provides an excellent example of a contemporary continental orogen formed by the accretion of terranes. Over 50 major terranes have been identified within the Western Cordillera, formed by the accretion of a wide variety of rock bodies to the North American continent, such as seamounts, oceanic islands, basaltic plateaus, and island arcs (see Section 7.3.1). The discovery and acceptance of the Cordilleran terranes has had a dramatic effect on our understanding and interpretation of the Earth's orogenic belts.

Source: Barber, T. 1985. A new concept of mountain building. *Geology Today* **14**, 116–121.

tectonics in the Western Cordillera of North America (Box 6.2) in the last 20 years or so has been highly influential in the reinterpretation of most orogenic belts.

There is increasing evidence that many ancient orogenic belts contain a complex collage of tectonic terranes consisting of micro-continents, island arcs and fragments of ocean floor all jammed together. Recognition of the boundaries between these different terranes, dating their respective collisions and determining the origin of each terrane is one of the most challenging areas of current research (see Section 7.3.1).

4. Transpression.   This involves deformation along a transform margin and the closure of any transform basins present (Figure 6.10). Transpression forms an important part of many orogenic episodes since there is usually a strong lateral component within most types of continental collision.

6.7   Summary of Key Points

- The Earth's surface consists of 7 large plates, 8 intermediate ones and over 20 small ones.
- There are three types of plate boundary: (1) divergent boundaries or spreading centres; (2) convergent boundaries; and (3) transform or conservative boundaries.
- Plate tectonics is modifed by plume tectonics, which involves upwelling of large bodies of material within the mantle. These superplumes may initiate and accelerate plate tectonic activity. In turn superplumes may in part owe their origin to the subduction of cold oceanic crust into the mantle.
- Plate tectonics explains the formation and geometry of many of the Earth's sedimentary basins today.
- The closure of sedimentary basins usually results in the formation of a mountain belt (orogenesis). There are four main types of orogenesis: (1) continental margin orogenesis; (2) continental collision; (3) collage collision; and (4) transpression.

6.8   Suggested Reading

There are many good accounts of plate tectonics in the literature, but both Dott and Batten (1988) and Duff (1992) provide particularly good summaries. Cox and Harte (1986) provide an excellent and detailed account of plate tectonics. The chapter on plate tectonics in Hallam (1990) provides a good historical background to the development of the plate tectonic model. This is complemented by good chapters on the development of the concept of plate tectonics and of the principles in Brown et al (1992). Barber (1985) and Harper (1998) provide excellent accounts of the processes of mountain building in the context of terranes.

References

Barber, T. 1985. A new concept of mountain building. *Geology Today* **14**, 116–121.

Brown, G.C., Hawkesworth, C.J. and Wilson, R.C.L. (Eds) 1992. *Understanding the Earth*. Cambridge University Press.

Cox, A. and Harte, R.B. 1986. *Plate Tectonics: How it Works*. Blackwell, Oxford.

Dott, R.H. and Batten, R.L. 1988. *Evolution of the Earth*, (Fourth Edition). McGraw-Hill, New York.

Duff, P. McL.D. (Ed.) 1993. *Holmes' Principles of Physical Geology*. (Fourth Edition). Chapman & Hall, London.

Hallam, A. 1990. *Great Geological Controversies*, (Second Edition). Oxford University Press, Oxford.

Harper, D.A.T. 1998. Interpreting orogenic belts: principles and examples. In Doyle, P. and Bennett, M.R. (Eds) *Unlocking the Stratigraphical Record*. John Wiley, Chichester, 491–524.

7

The Stratigraphical Tool Kit:
Case Studies

In this chapter we provide a series of extended examples or case studies which illustrate the application of the stratigraphical tool kit introduced in the previous chapters. The intention is not to illustrate every principle but simply to give an insight into some of the methods by which stratigraphers build up a picture of events in Earth history. For convenience our examples have been grouped under three headings: (1) establishing and dating the sequence; (2) interpretation of the sequence and; (3) the application of stratigraphy to plate tectonics.

7.1 Establishing and Dating the Rock Sequence

Within this section there are two case studies. The first examines the work which unravelled the chronology of the Lewisian rocks of the Scottish Precambrian. It demonstrates the importance of cross-cutting relationships in determining the relative order of rock units within an area. The second case study illustrates some of the concepts of sequence stratigraphy and discusses its application in field studies.

7.1.1 Unravelling Precambrian Chronology in Scotland

(*Sutton and Watson 1951; Giletti et al 1961*)

1. Introduction
The metamorphic Lewisian complex of the Scottish Highlands contains some of the oldest rocks in Britain, dating from the Archaean (Figure 7.1). The Lewisian forms

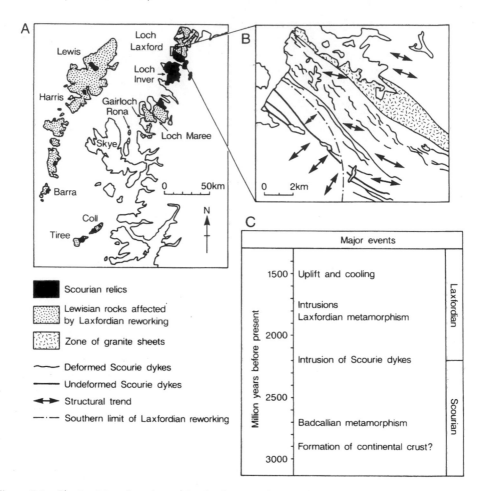

Figure 7.1   *The Lewisian of north-west Scotland.* **A:** *Distribution of Laxfordian reworking in the Lewisian rocks.* **B:** *Map of the Scourian to Laxfordian boundary in the Loch Laxford area, showing the contrast between the deformed and undeformed Scourie dykes.* **C:** *Chronology of Lewisian events.* [Modified from: Anderton et al (1979)* A Dynamic Stratigraphy of the British Isles, *Allen & Unwin, Figures 2.2 & 2.3, pp. 16 & 17]*

the basement for much of the younger rock cover in Scotland. The complex has been subject to several episodes of intense metamorphism and as a consequence developing a stratigraphy for it has proved difficult.

The Lewisian is composed of gneisses which vary in composition from ultrabasic to acidic. Early investigations of the complex were undertaken by members of the Geological Survey of Great Britain in the early part of the twentieth century, who established a basic chronological framework. The Lewisian gneisses were observed to have been intruded by a younger group of basic igneous intrusions (dykes) which were later metamorphosed into basic gneisses. This relative sequence of rocks and events within the Lewisian was not challenged until the work of Sutton and Watson (1951). Through detailed field observations and the interpretation of simple cross-cutting relationships Sutton and Watson were able to show that two phases of metamorphism had occurred: one before the intrusion of the dykes and one after. They related their metamorphic episodes directly to orogenic events and suggested that the igneous intrusions represented a punctuation mark between them (Figure 7.1). However, what Sutton and Watson lacked was an independent means of confirming the time equivalence of the igneous intrusions: did they all form at the same time? They also had no clear idea of the time scale over which these orogenic episodes took place. This was provided 10 years later by the application of radiometric dating. Giletti et al (1961) represents one of the earliest attempts at applying radiometric dating techniques to interpret complex field relationships. Giletti and his colleagues were able to put firm constraints on the timing of the metamorphic events.

2. Cross-cutting relationships – Sutton and Watson (1951)

From a basis of field observation in the Loch Torridon and Scourie areas of the north west Highlands of Scotland Sutton and Watson (1951) were able to subdivide the Lewisian gneisses on the basis of lithology. The relative chronology of these lithological units was established on the basis of the intrusion, tectonic and metamorphic history of a series of north-west to south-east trending dolerite dykes, known as the Scourie dykes. They recognised that the gneisses exposed between Gruinard Bay and Scourie were of high metamorphic grade (granulite-facies) and possessed folded structures which had a north-east trend. Sutton and Watson demonstrated that these rocks were traversed by a swarm of north-west to south-east trending dolerite dykes which were unaffected by this metamorphism (Figure 7.1). The dykes displayed chilled contacts with the gneisses, and this clearly illustrated that the dykes were intruded into the gneiss long after it had cooled. Clearly, therefore, the intrusion of the dykes post-dated the metamorphism of these gneisses. Farther to the north and to the south of this area, the gneisses were of different composition, rich in the mineral amphibole, typical of lower grade metamorphism. The structural grain in these rocks trended to the north-west (Figure 7.1). These rocks had also been intruded by the dolerite dykes, but in this case the dykes had also suffered metamorphism and deformation (Figures 7.1 and 7.2). On this basis, Sutton and Watson developed their model of two major episodes of metamorphism in the development of the Lewisian complex: an older phase, the Scourian, and a younger phase, the Laxfordian, which reworked the Scourian gneisses and the intervening dykes. They regarded the intrusion of the Scourie dykes as time significant, representing an event between the two episodes of metamorphism: 'the period of dolerite intrusion may therefore be taken as a fixed point in the history of the north-west Highlands' (Sutton and Watson 1951, p. 292).

3. Unravelling the ages – Giletti et al (1961)

Giletti et al (1961) were able to place firm constraints on the timing of the metamorphic episodes recognised by Sutton and Watson (1951) in the field. They used Rubidium – Strontium dating of micas and potassium feldspars in the gneisses. The metamorphism of the Scourian complex was dated to between 2460 and 1900 Ma ago. The Laxfordian was dated in the range of 1160 to 1620 Ma ago. Giletti et al confirmed that these dates did not indicate one long extended period of metamorphism but

Figure 7.2   *Photograph of Scourie dykes deformed during the Laxfordian orogeny, from the Isle of Barra, north-west Scotland [Photograph: BGS]*

rather two separate geological episodes between which there was a period of dyke intrusion (Figure 7.1). This work supported and enhanced the original conclusions of Sutton and Watson.

4. Comment

This case study demonstrates the importance of cross-cutting relationships in determining the relative chronology of events within an area. In particular it illustrates the application of these tools to structurally complex, high grade metamorphic terrains and shows that the application of simple stratigraphical techniques can provide important information in determining the relative sequence of geological events within an area. Further information about this story can be obtained in the excellent review by Rodgers and Pankhurst (1993).

5. References

Giletti, B.J., Moorbath, S. and Lambert, R.St.J. 1961. A geochronological study of the metamorphic complexes of the Scottish Highlands. *Quarterly Journal of the Geological Society of London* **117**, 233–272.

Rogers, G. and Pankhurst, R.J. 1993. Unravelling dates through the ages: geochronology of the Scottish metamorphic complexes. *Journal of the Geological Society of London* **150**, 447–464.

Sutton, J. and Watson, J. 1951. The pre-Torridonian metamorphic history of the Loch Torridon and Scourie areas in the north-west Highlands and its bearing on the chronological classification of the Lewisian. *Quarterly Journal of the Geological Society of London* **56**, 241–295.

7.1.2   The Application of Sequence Stratigraphy in Field Studies: an Example from the Early Palaeozoic Welsh Basin

(*Woodcock 1990*)

1. Introduction

The concept of sequence stratigraphy was developed for use with subsurface borehole and seismic data. However, the concept has recently been applied to onshore basins using outcrop data, and it has been found to provide a valuable tool in determining the regional history of a sedimentary basin. A good example of this type of application is provided by Woodcock's (1990) work on the Early Palaeozoic Welsh Basin. During the Palaeozoic much of the area which we now know as Wales was a deep sedimentary basin along the southern margin of an ocean called the Iapetus Ocean. In his paper Woodcock erected a new stratigraphy for this basin based on the principles of sequence stratigraphy and using this he established a chronostratigraphical division of these rocks.

2. Relevant principles of sequence stratigraphy

As we discussed in Section 5.1.2 a depositional sequence is a three-dimensional rock body, within a sedimentary basin. The boundaries of a sequence are defined for the most part by unconformities (chronostratigraphical breaks). It must, however, be remembered that individual sequence boundaries may grade from a clear unconformity at the margin of a basin, through disconformity, to conformity at the basin centre (Figure 5.10). The presence of basin-wide unconformities can be used to group 'sequences' into megasequences.

A tentative guide to the sequence stratigraphy of a basin can be obtained by calculating a rock preservation curve. A rock preservation curve for an area shows the amount of rock preserved at successive times as a percentage of the potential maximum. The curve is computed from stratigraphical logs, derived from bore holes or rock sections, and the method by which it is calculated is shown in Figure 7.3. On a rock preservation curve megasequence boundaries are recorded as 0 per cent lows on the curve, representing zero preservation across the basin (i.e. no rock deposition anywhere in the basin). The remaining lows represent unconformities which are less than basin-wide and are therefore sequence boundaries.

3. Application to the Welsh Basin

Woodcock (1990) drew up a rock preservation curve for the Welsh Basin in the west of Britain, on the basis of numerous stratigraphical logs (Figure 7.4). On this curve he was able to recognise four points at which there was zero preservation, which he used to define three megasequences. He then established which lithostratigraphical units belonged to which megasequence by reference to the original stratigraphical logs. On this basis he was able to set up three supergroups – the Dyfed, Gwynedd and Powys supergroups – corresponding to the three megasequences present. A further 19 sequence boundaries were also identified within the curve, giving 18 distinct depositional sequences. In this way Woodcock was able to establish a basin-wide stratigraphy for the Welsh Basin, in which individual records or sections could be placed.

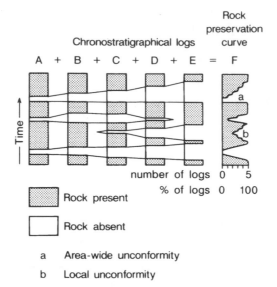

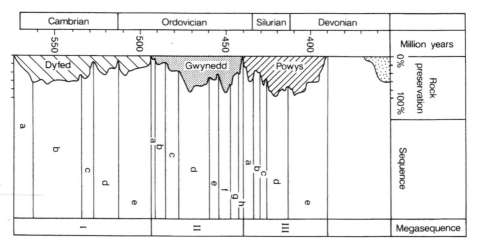

Figure 7.3   Construction of a rock preservation curve (F) by totalling the amount of rock present in a series of chronostratigraphical logs (A–E) located throughout a depositional basin. [Modified from: Woodcock (1990) Journal of the Geological Society of London **147**, Figure 2, p. 538]

Figure 7.4   Rock preservation curve for the Early Palaeozoic Welsh Basin showing the main sequences and megasequences identified (% rock volume through time). [Modified from: Woodcock (1990) Journal of the Geological Society of London **147**, Figure 3b, p. 539]

Woodcock went on to interpret the sequence stratigraphy he had established. He recognised that the majority of the proposed sequence boundaries in the Welsh Basin were probably caused by processes other than just purely eustatic sea-level change. This conclusion is of considerable importance since traditionally sequence stratigraphies are simply interpreted in terms of eustatic sea level variation. Woodcock (1990) and others have suggested that sequence stratigraphies are controlled by tectonism, sedimentation and eustatic sea level variation. Consequently, sea level curves deduced from sequence stratigraphy may not necessarily give a precise picture of eustatic sea level.

4. Comment

This case study illustrates two processes of investigation. First, detailed analysis of the stratigraphical record to identify stratigraphical breaks. Second, the application of this data in order to construct a chronostratigraphy, which provides the temporal framework with which to group lithostratigraphical units.

5. Reference

Woodcock, N.H. 1990. Sequence stratigraphy of the Palaeozoic Welsh Basin. *Journal of the Geological Society of London* **147**, 537–547.

7.2   Interpreting the Rock Sequence

Within this section there are two case studies which examine how rock units can be interpreted in terms of palaeogeography, palaeoclimate and palaeoecology. The first case study deals with a facies interpretation of the palaeoenvironments represented by the Permian rocks of the North Sea. The second case study deals with the application of fossil assemblages (biofacies) in determining palaeo-shorelines.

7.2.1   Permian Redbeds of North-West Europe: an Example of Palaeoenvironmental Reconstruction

(*Glennie 1972*)

1. Introduction

The Permian redbeds of north-west Europe, known as the *Rotliegendes*, are clastic sediments originally deposited on land, in arid or semi-arid conditions. Their depositional environment is primarily a function of the development of a semi-arid environment which resulted from the construction of a large supercontinent at the end of the Palaeozoic (Pangaea). These sediments occur extensively beneath the North Sea and are host to much of the natural gas present in this area. As a consequence they have been the subject of considerable research. Glennie (1972) provided a remarkable survey of the facies present within the *Rotliegendes* and his contribution is considered by many to be a classic example of facies analysis. His approach was first, to review the sediments found in modern desert environments and second, to use these modern analogues to interpret the sediment cores taken from boreholes drilled beneath the North Sea.

2. Sediments of Modern Deserts

Glennie (1972) identified two main types of sediments deposited in desert environments today: those deposited by water, and those deposited by wind.

1. Water-lain sediments – Glennie recognised that there are two main types of water-lain sediments present in desert environments, distinct fluvial sediments and temporary desert lake sediments.

    A. Desert fluvial sediments. Although they are often infrequent, heavy rainstorms do occur in deserts and because of the absence of vegetation they are able to move large amounts of debris and cut valleys known as wadis. The ephemeral rivers produced by these storms often dry out before reaching the sea or inland lake and therefore dump their sediment load as an unsorted conglomerate within their wadis.

    B. Temporary desert lakes and inland sabkhas. Temporary lakes may form in the centre of a basin with inland drainage, or where sand dunes block a wadi channel. As the water evaporates, salts are deposited and may form a crust of halite or gypsum salt over the sediment surface. Flat areas of salt-encrusted sand, silt and clay are known as sabkhas. Another phenomenon common in this type of environment are sand dykes. A layer of silt and clay is commonly deposited on the floor of a temporary lake as it dries out. This clay layer will become desiccated as the lake dries up and cracks will form. These cracks frequently become filled with a slurry of wet sand from below the dry surface and the structures produced are known as sand dykes. Alternatively the desiccation cracks may become filled by wind blown sand from above. Adhesion ripples may also form in this type of environment. When dry sand is blown across a wet sediment surface, some grains stick to the surface on impact. If this process continues an irregular, blistered or ripple-like surface may develop (adhesion ripples).

Figure 7.5   *Photograph of a linear sand dune from the Namibian desert. [Photograph: J.E. Bullard]*

2. Wind deposits (aeolian) – Glennie recognised that wind blown sand is deposited either in continuous sheets or as discrete dunes (Figure 7.5). Internally sand dunes contain distinct sedimentary structures (cross bedding) which not only allows the type of dune to be inferred but also provides evidence of the palaeo-wind direction. Aeolian sediments may occur in association with water lain sediments. The fluvial sands of ephemeral rivers form an important source of wind blown sand and during the long intervals between rainstorms when the wadi channels are dry the fluvial sands may be reworked by the wind into small dunes. When the wadi is next in flood these aeolian sands are often removed or 'flushed out'.

3. The Sediments of the Upper Rotliegendes of the North Sea

Figure 7.6 presents two generalised sedimentary sections taken from the Upper Rotliegendes rocks beneath the North Sea. They are based on numerous sediment cores drilled during the exploration for natural gas and oil. Glennie (1972) interpreted the sedimentary structures present within these terrestial redbeds on the basis of the modern desert sediments reviewed in the first part of his paper. From an analysis of the cross bedding present within the aeolian sands Glennie (1972) was also able to conclude that both barchan and sief dunes were present and that the palaeo-wind blew predominantly from east to west.

By looking at the spatial distribution of sediment facies beneath the North Sea in sediment cores Glennie (1972) was able to produce a conceptual cross-section or model of the facies distribution within the Upper Rotliegendes rocks beneath the North Sea and the Netherlands (Figure 7.7). In the south, this cross-section shows wadi sediments spreading out from the Hercynian mountains. These sediments pass into an extensive dune field, beyond which there was a broad desert lake subject to periodic desiccation in the north.

4. Comments

The work of Glennie (1972) provides a classic example of the use of facies analysis to give a picture of an ancient period in Earth history. The application of modern analogues emphasises the importance of uniformitarianism in facies analysis. It also shows how a palaeogeographical model of the distribution of the desert environments may be drawn up from borehole and outcrop data. This story has been reviewed and updated by George and Berry (1993).

5. References

George, G.T. & Berry, J.K. 1993. A new lithostratigraphy and depositional model for the Upper Rotliegend of the UK Sector of the Southern North Sea. In: North, C.P. and Prosser, D.J. (Eds) *Characterization of Fluvial and Aeolian Reservoirs*. Geological Society Special Publication No. 73, 291–319.

Glennie, K.W. 1972. Permian Rotliegendes of north-west Europe interpreted in the light of modern desert sedimentation studies. *Bulletin of the American Association of Petroleum Geologists* **56**, 1047–1071.

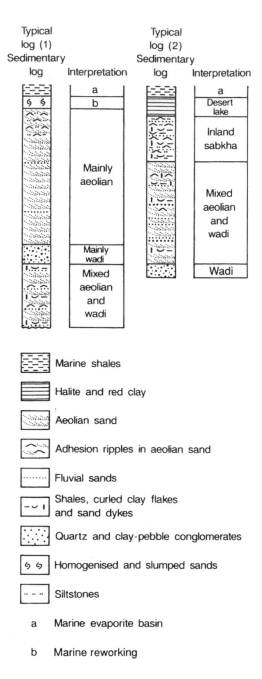

Marine shales

Halite and red clay

Aeolian sand

Adhesion ripples in aeolian sand

Fluvial sands

Shales, curled clay flakes
and sand dykes

Quartz and clay-pebble conglomerates

Homogenised and slumped sands

Siltstones

a   Marine evaporite basin

b   Marine reworking

Figure 7.6 *Two idealised sedimentary logs from boreholes through the Rotliegendes of the North Sea. The facies interpretation is shown to the right of each log. [Modified from: Glennie (1972) Bulletin of the American Association of Petroleum Geologists **56**, Figure 9, p. 1055]*

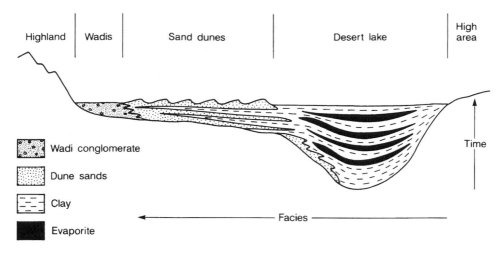

| Highland | Wadis | Sand dunes | Desert lake | High area |

Wadi conglomerate

Dune sands

Clay

Evaporite

Time

← ————————————— Facies ————————————— →

Figure 7.7 *Environments and facies associations of the Rotliegendes of the North Sea, along a transect running north from the Hercynian mountains located in the southern North Sea [Modified from: Glennie (1972) Bulletin of the American Association of Petroleum Geologists* **56***, Figure 17, p. 1067]*

7.2.3 Silurian Coastlines

(*Ziegler 1965*)

1. Introduction

Living organisms can be grouped into communities of two or more species which occupy the same environment or 'ecospace'. The balance of different organisms in a given environment depends on their relative success, which is ultimately determined by the physical environment. Environmental variables such as the amount of available oxygen, water temperature and rate of sedimentation control the ultimate fate of a community of organisms. Building up a clear picture of fossil communities in the geological record is difficult since few soft-bodied organisms are preserved and because accumulation of fossils usually takes place over a considerable time period, during which a community may change. Despite these limitations it is possible to reconstruct ancient fossil communities from recurrent fossil assemblages in some locations, and consequently one can deduce the environmental parameters which limited them. Consequently the distribution of fossil communities should provide important information with which to reconstruct ancient palaeogeographies. The work of Ziegler (1965) provides a very clear illustration of the application of this idea in the Lower Palaeozoic rocks of Wales.

Ziegler's (1965) work is based on the detailed observation of fossil assemblages (biofacies) present in the rocks of the Lower Silurian (Llandovery) in Wales and the Welsh Borders. The area was located on the southern margin of a contracting ocean, the Iapetus Ocean, during the Silurian. These rocks consist of turbidite and dark mud facies in the north-west and sandier facies to the south-east. Ziegler found that there were five recurring assemblages of bottom-dwelling organisms. The content of these assemblages was dominated by brachiopods. The detailed nature of each assemblage was described in detail in a later paper (Ziegler et al 1968). Brachiopods are marine, shelled organisms with a delicate feeding mechanism and are not widely distributed today. In the Early Palaeozoic, however, they were amongst the commonest type of sea floor organism. Ziegler recognised each of his five assemblages or communities on the basis of common species, as well as on the relative abundance of individual species within the community (Figure 7.8). In the 1920s the Danish ecologist Petersen had discovered that it was possible to map the relative distribution of marine communities in the shallow seas of today. He determined that they form distinct patterns reflecting the distribution of marine environments. On this basis Ziegler (1965) argued that the recurrent set of fossil communities recognised in the rocks of the Llandovery in Wales reflected the distribution of marine environments along the southern margin of the Iapetus Ocean.

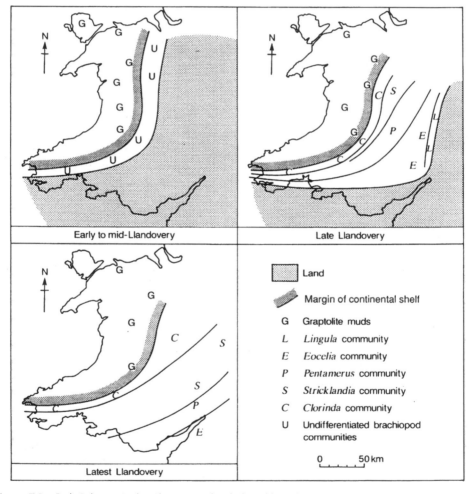

Figure 7.8 *Early Palaeozoic shorelines in Wales deduced from biofacies. The shorelines plotted on the maps were constructed from the distribution of fossil assemblages. The maps depict a progressive marine transgression during the Llandovery. [Modified from: Ziegler (1965) Nature 207, Figures 1–3, pp. 270–271]*

Ziegler mapped the geographical distribution of his five fossil communities in the Llandovery outcrop of Wales and the Welsh Borders. From this it became apparent that the five communites fell into a number of arcuate, concentric zones at any point in time and that their relative position remained constant. For example, if one was to move progressively westwards across the area the following communities would be encountered: *Lingula, Eocelia, Pentamerus, Stricklandia* and *Clorinda*, before one moved into the basinal muds in which there were few bottom living organisms. The *Lingula* community borders the upstanding Precambrian massif of the Malvern Hills in central England and Zeigler (1965) argued that the succession of fossil communities represented a progression offshore: the communities changing with increasing water depth and decreasing water turbulence. In this way the arcuate pattern of communities could be interpreted in terms of an ancient shoreline (Figure 7.8).

Zeigler (1965) developed this concept further by mapping the distribution of these five fossil assemblages through time as well as space. In this way he was able to track the movement of the coastline inshore during the Llandovery. The vertical changes in the type of communities present with the rock sequence reflects the progressive migration of deeper water coastal zones onshore (Figure 7.8).

2. Comment

This case study illustrates the application of uniformitarian principles to the interpretation of biofacies. Given that fossil communities can be recognised the interpretation of the mode of life of the animals within a biofacies enables a reconstruction of the environment in which they lived and in some cases even of the palaeogeography.

3. References

Ziegler, A.M. 1965. Silurian marine communities and their environmental significance. *Nature*, **207**, 270–272.

Ziegler, A.M., Cocks, L.R.M. and Bambach, R.K. 1968. The composition and structure of Lower Silurian marine communities. *Lethaia* **1**, 1–27.

7.3   The Application of Stratigraphy to Plate Tectonics

Plate tectonics has provided an important model with which to interpret the stratigraphical sequence. However, the application of simple stratigraphical tools has also played an important role in refining the plate tectonic model. The case study chosen in this section tells the story of the recognition and development of the concept of terranes. As outlined in Box 6.2 terranes are produced by the accretion of different 'slivers' of crust onto a continental margin during a process known as collage tectonics. The initial recognition of terranes in the North American Cordilleran mountains was based upon simple stratigraphical analysis of the Western Cordillera, which displays widely differing stratigraphies and fossil faunas in close juxtaposition, separated by faults.

7.3.1 Allochthonous Terranes

(Monger and Ross 1971; Irwin 1960, 1972; Coney et al 1980; Tipper 1982)

1. Introduction

Tectonic belts are linear belts of deformed crust. Before the advent of plate tectonic theory these belts were interpreted as geosynclines or regional downwarps in which sediment accumulated and was deformed. With the advent of plate tectonic theory these linear belts were reinterpreted in terms of continental collison and the Wilson Cycle (Figure 6.5). This model of mountain building has developed further in recent years with the recognition of allochthonous terranes within many tectonic belts. Allochthonous terranes are fault-bounded tracts of land each with a distinct strati-graphical sequence and fauna. Today many mountain belts are interpreted in terms of terranes and are believed to have formed by the accretion of a variety of different 'slivers' of crust (terranes) which are accreted either between colliding continents or along the edge of a continent to produce mountain belts (Figure 6.11). The initial recognition of allochthonous terranes owes much to the field observation of different stratigraphies and therefore geological histories as well as the recognition of unusual or foreign faunas. The concept of allochthonous terranes was developed in the North American Western Cordilleran Mountains, where in excess of 50 terranes have been recognised. This work and the development of the terrane concept has had a pro-found effect on the interpretation of tectonic belts.

2. Recognition of Disjunct Faunas and the Discovery of Terranes

Monger and Ross (1971) are palaeontologists who were then studying the distribution of the fossil remains of large single-celled organisms (foraminifera) called fusulinids. Fusulinids lived on the sea floor and were common in the Late Palaeozoic, becoming extinct during the Late Permian. In studying the Permian rocks of British Columbia, Monger and Ross recognised that in the area around Cache Creek the fusulinid fauna was of European aspect, with the species present common only in Permian marine rocks of the European Alpine belt. This contrasted with areas to the west and east of Cache Creek which contained faunas of North American affinity. The implication of this was that the Cache Creek region was actually displaced and allochthonous to the region. This work revived interest in observations made much earlier by Irwin (1960).

Irwin (1960) had recognised that the Klamath Mountains in Northern California could be divided into four belts on the basis that each belt had a distinct fauna and that the nature of the stratigraphy in each belt was different. In his later work Irwin showed that these belts were fault bounded and that each sequence was therefore a separate tectono-stratigraphical entity brought together by tectonic activity, but not having been originally in close juxtaposition (Irwin 1972). This led to Irwin's (1972) definition of a terrane as 'an association of geological features, such as stratigraphical sequences, palaeontology, intrusive rocks, mineral deposits and tectonic history which differ from those of an adjacent terrane'.

3. Wider Implications

Coney et al (1980) reviewed the body of evidence about allochthonous terranes that had been growing from the initial palaeontological and stratigraphical data. They recognised in excess of 50 terranes accreted onto the ancient rocks of the North American craton (Figure 7.9). They recognised terranes on the basis of their internal homogeneity and the continuity of stratigraphy, tectonic style and geological history within their boundaries. The terrane boundaries separate distinct stratigraphies within the terranes and juxtapose different faunas. In this way the recognition of

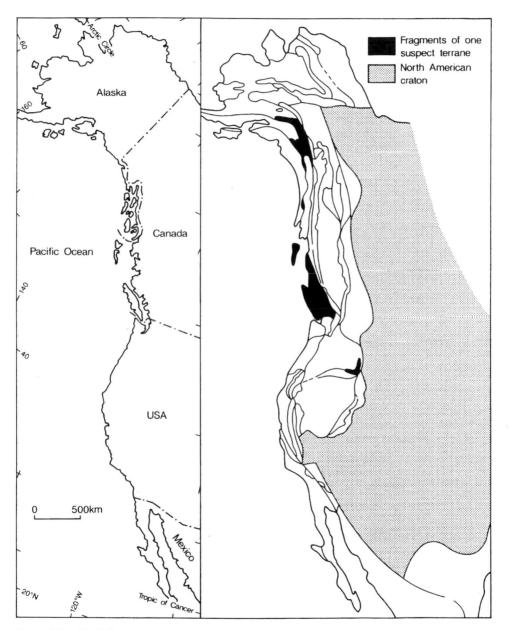

Figure 7.9 *Allochthonous terranes within the North American Cordillera. The fragments of one suspect terrane are picked out in black. [Modified from: Coney et al (1980) Nature* **288**, *Figure 1, p. 330]*

individual terranes is based solely on the stratigraphical character of the terrane. A terrane is first recognised on the basis of a stratigraphical history unlike that of an adjacent terrane and distinct enough not to be a result of a facies change. Many of the terranes recognised by Coney et al (1980) have stratigraphical histories indicative of an oceanic setting, with oceanic sediments and basaltic lavas, while others seem to represent fragments of volcanic arcs or small continental fragments. In turn these are distinguished from the Palaeozoic and pre-Palaeozoic stable craton of the North American interior. The Late Mesozoic history of these terranes shows that they were accreted during this period onto the Cordilleran margin. The date at which each terrane docked with the North American craton can be determined by an analysis of cross-cutting relationships and the dating of igneous bodies of rock associated with the docking of a particular terrane.

Coney et al (1980) postulated that the origin of the terranes was connected with the closure of the Pacific Ocean, which was dominant in the Late Palaeozoic to Early Mesozoic when most continental crust was part of the large landmass of Pangaea. Island arcs and fragments of oceanic crust were swept from the west against the cratonic margin and moved laterally along it to produce the mosaic of terranes that we see today.

4. The Importance of Biogeography to Terrane Analysis

Tipper (1984) demonstrated that the distribution of fossil organisms is a strong tool in the interpretation of the stratigraphical record of the Cordilleran terranes. He found that the ammonite faunas of the Canadian Cordillera could be used to show the relative lateral displacement of terranes both before and after they docked with the North American continent (Figure 7.10).

It is well-established that in common with many other Jurassic marine organisms that ammonite species are organised into distinct latitudinal belts which may reflect the presence of the main climatic belts on the Earth's surface. In the later part of the Early Jurassic the ammonites became segregated into latitudinal faunas, usually referred to as the Boreal (northern) and Tethyan (southern) realms. These realms can be identified in the rocks deposited on the flooded craton at these times as well as in the Canadian terranes of Wrangellia, Stikinia and Quesnellia. It is clear that these three terranes had differing stratigraphical histories until the Middle Jurassic, and that in the Late Jurassic all three terranes had moved northwards along faults which bounded each terrane. This was apparent because the latitudinal position of the faunal realms in the three terranes has been offset and is at odds with the distribution of the faunal realms on the stable craton. Docking and the lateral displacement of these three terranes was completed by the Middle Jurassic, since they are stratigraphically and faunally similar thereafter.

5. Comment

The discovery of terranes has had a profound effect on how we view tectonic belts. Terranes are identified on stratigraphical and faunal evidence within many of the Earth's tectonic belts, and their recognition increased our understanding of tectonic processes. The case history illustrates that stratigraphical analysis of lithostratigraphy, biostratigraphy and facies are powerful tools in plate tectonic modelling.

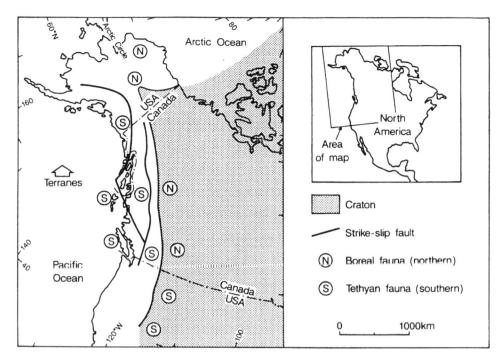

Figure 7.10 *The use of faunal realms to deduce the lateral movement of terranes within the Canadian Cordillera. The boundary between the Boreal and Tethyan faunal realms has been displaced northwards in the area of accreted terranes [Modified from: Tipper (1984) In: Westermann (Ed.) Jurassic Biochronology and Biogeography, The Geological Association of Canada, Special Paper 27, Figure 3, p. 115]*

6. References

Coney, P.J., Jones, D.I. and Monger, J.W.H. 1980. Cordilleran suspect terranes. *Nature* **288**, 329–333.

Irwin, W.P. 1960. Geological reconnaissance of the northern Coast Ranges and Klamath Mountains, California. *Californian Division of Mines Bulletin* **179**.

Irwin, W.P. 1972. Terranes of the western Paleozoic and Triassic belt in the southern Klamath Mountains, California. *US Geological Survey Professional Paper* **800**-C, 103–111.

Monger, J.W.H. and Ross, C.A. 1971. Distribution of fusulinaceans in the western Canadian Cordilliera. *Canadian Journal of Earth Sciences* **8**, 259–278.

Tipper, H.W. 1984. The allochthonous Jurassic–Lower Cretaceous terranes of the Canadian Cordilliera and their relation to correlative strata of the North American craton. In: Westermann, G.E.G. (Ed.). *Jurassic–Cretaceous biochronology and paleogeography of North America*. Geological Association of Canada Special Paper No. 27, 113–120.

7.3 Summary of Key Points

In this chapter we have brought together several case histories to illustrate the basis of stratigraphical procedure, and the interpretation of the rock succession in terms of a sequence of events in Earth history. These case studies all illustrate that stratigraphy involves three basic elements: observation, description and interpretation. In establishing the rock succession the stratigrapher observes and describes the rock units present. When dating the sequence, using either relative or absolute techniques, the stratigrapher establishes a chronostratigraphical framework for the units he or she has described. The description of units is vital for their interpretation. Interpretation of rock units in terms of ancient environments, ecologies and geographies is under-pinned by the principle of uniformitarianism. Most geologists accept that plate tectonics is the unifying concept in geology, but as the last case study has shown, the plate tectonic model is frequently refined. In this case allochthonous terranes were recognised through the simple observation, description and interpretation of stratigraphical sequences. The implications of these simple observations have caused the plate tectonic model to be refined which has in turn refined our interpretation of tectonic belts.

8

Summary of Part One: The Stratigraphical Tool Kit

In the first part of this book we have examined the stratigraphical tools necessary to fulfil the main aim of stratigraphy, which is the interpretation of sequences of rock as a series of events in the evolution of the Earth. In order to do this there are three tasks which must be undertaken: (1) to establish the sequence of rock units present; (2) to work out the chronological order of those rock units; and (3) to interpret the environment in which each unit was deposited or formed, so that a record of the changing environment can be established. Since the 1960s plate tectonics has provided the broad theoretical framework with which to view the global development of the stratigraphical record and therefore a fourth task is to determine the global perspective with a stratigraphical sequence in the light of plate tectonic theory. Part One has introduced the basic tools which will enable you to carry out these tasks and fulfil the aim of stratigraphy. A recap of these tools is provided below.

Actualism—the modern interpretation of uniformitarianism—recognises that the laws of nature have remained unchanged throughout most of geological time, although the rates at which such processes operate may have varied. The concept of actualism underpins our understanding and interpretation of the geological record. It is the most fundamental principle in stratigraphy and the most important element of the stratigraphical tool kit.

The special component of geology which sets it apart from almost all other sciences is that of time. It is important to determine the relative chronology or timing of the sequence of rock units. In bedded sequences, superposition determines that the lowest layer in an undeformed sequence is the oldest. In non-bedded sequences, the

cross-cutting of rock masses by faults or igneous bodies, for example, helps to determine the relative chronology. Absolute geological time may be determined by the
radioactive decay of unstable isotopes, but in practice the relative order of events is
sufficient for most geologists. The relative order may be determined by the position
of evolving fossils or through the recognition of instantaneous events. The
Chronostratigraphical Scale is important as it provides a global standard for correlation of geographically or environmentally different sequences.

The rock record is interpreted for the most part through the application of
actualism and the recognition that the process which has operated can be determined
from the characteristics of the product. In this way, the sum total of the characteristics
of a rock body, called its facies, can be used to interpret the environment prevailing
during its formation or deposition. The recognition of facies logically leads to the
construction of a palaeogeography, each facies corresponding to an environment
which may be plotted on a map to represent the geography of an area at any one
time. Climate has a profound effect on sedimentation: the prevailing palaeoclimatic
conditions at the time of deposition can therefore be inferred from the characteristics
of many sedimentary rocks. Fossils too provide an important insight into the
development of an environment at any one time and palaeoecologies can be reconstructed.

Part One of this book is therefore about the ways and means of obtaining detailed
information from the stratigraphical record. In most cases this means the collection of
local information from local rock sections. The work of countless geologists has
contributed in this way to the recognition of a global pattern in the rock record in
both space and time. Part Two illustrates the record of global change determined
from the tools outlined in the previous pages and shows these changes to have
interacted to produce the Earth's stratigraphical record.

Part Two
THE PATTERN OF EARTH HISTORY

9

Introduction to Part Two

As we have shown in Part One, stratigraphy is the interpretation of rock successions in terms of sequences of events in Earth history. Each stratigraphical unit contains a number of clues to the nature of the environment in which it was formed. By piecing together these clues it is possible to determine the nature of the changing environments of the Earth through time. Stratigraphy therefore provides the temporal framework within which we can compare ancient environments and study the way in which the Earth has changed.

The traditional approach to stratigraphy is to build up a pattern of Earth history brick-by-brick, layer-by-layer. Only when all the layers are in place can the global pattern and large scale controls on the development of the stratigraphical record be appreciated. In this part of the book we shall attempt to provide an alternative approach in which we shall build up a framework of the global controls on the development of the stratigraphical record and their variation through geological time. We shall do this before illustrating how this framework can be used to view the geological development of one small part of the Earth's crust, in this case the North Atlantic region.

What are the global or large scale controls on the stratigraphical record? Today, the Earth's physical environment is the product of the interaction of four variables: plate tectonics, climate, sea level and the biosphere. If one applies the principle of uniformitarianism then the ancient environments deduced from each stratigraphical unit can also be interpreted in terms of these four variables. Together they give a record of Earth history. These four variables control the nature of the depositional environment and together have created the stratigraphical record. Within these variables there is a distinct hierarchy of importance (Figure 9.1). Plate tectonics is at the head of this hierarchy and is fundamental because it controls the formation and

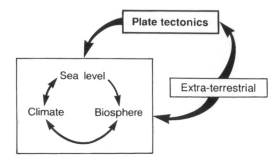

Figure 9.1   *Flow diagram showing the interaction of the key variables which determine the nature of the physical environment of the Earth*

development of depositional basins within which the sediments, which comprise much of the stratigraphical record, can accumulate. Plate tectonics also determines the production and distribution of igneous and metamorphic rocks and the nature of the tectonic processes which deform the stratigraphical record. In contrast, climate, sea level and the biosphere, although having a profound influence on sedimentary environments, only control the nature of the sedimentary facies preserved in the depositional basins that are created by plate tectonics. These three variables are therefore strongly dependent on plate tectonics and are also closely interrelated with one another (Figure 9.1). In this way, the interaction of the four variables is responsible for the production of the Earth's stratigraphical record. It is, however, important to note that the Earth may also have experienced extra-terrestrial influences, particularly the impact of meteorites and other similar bodies. Extra-terrestrial impact has frequently been invoked to explain changes within the biosphere, climate and also in plate tectonic activity, although in many cases these hypotheses remain speculative.

In Part Two, Chapter 10 explores the way in which plate tectonics, sea level, climate and the biosphere have varied globally throughout Earth history, each with their own independent but interrelated patterns of temporal change. This is only possible because of the detailed investigations of generations of 'stratigraphical detectives' who have pieced together the sequence of events and interpreted their significance. In the remaining chapters of this book the record of global change is examined with reference to the stratigraphical record of one part of the Earth's crust, the North Atlantic region, comprising northern Europe and North America.

10

The Stratigraphical Record and Global Rhythm

The stratigraphical record can be interpreted in terms of the interaction of four variables: plate tectonics, climate, sea level and the biosphere. These four variables combine to determine the nature of the rocks deposited at any point on the surface of the Earth. Throughout Earth history these four variables have varied systematically through time, each with a different rhythm. In this chapter the rhythm of change through time of each of these four variables is explored.

10.1 Plate Tectonics

Plate tectonics controls the evolution, shape and closure of sedimentary basins and is therefore probably the most important control on the development of the stratigraphical record. It is also significant in that it drives the rock-cycle, creating igneous rocks as a primary source of clastic sediments, and producing metamorphic zones. Ultimately, plate tectonics is the means by which the stratigraphical record is written, preserving the leaves in our book of history, while by contrast variables such as sea level and climate simply determine the character of that record, or to continue the analogy, the colour of the ink with which it is written. As an illustration of this, deposition may occur widely, but the preservation of sediments is largely controlled by the availability of basins in which they can accumulate. Without such basins there would be little sediment preserved and our record of the Earth's environments through time as represented by the stratigraphical record would be poor. The formation and subsidence of sedimentary basins is therefore critical to any interpretation of the stratigraphical record, and this theme is explored in this section.

10.1.1   Plate Tectonics Through Time

Since the introduction of the plate tectonic model in the 1960s it has been established that plate tectonic processes have been operative throughout at least the Phanerozoic, and probably existed in some form during the later part of the Precambrian. Plate tectonics has provided an important key to understanding the Earth's stratigraphical record, but how far back can the model be applied?

The Earth's interior has been cooling since its formation some 4600 million years ago (Figure 10.1) and with it geotectonic processes have changed. What is emerging from recent work is a subtle shift in process from plume dominated tectonics to plate tectonics throughout the Precambrian as the mantle cooled and the thermal stratification within it became less pronounced. Currently there is active debate and discussion of geotectonic models for the Early Archaean and Proterozoic, and as these are still largely theoretical concepts, what follows is just one possible scenario.

The Earth's first crust probably formed on an ocean of magma some time after 4500 Ma. Rapid convection in the upper mantle led to recycling of this early crust, but the exact nature of the plate boundaries and in particular the geometry of the associated subduction zones is uncertain. For example, in some scenarios subduction actually involves the simultaneous subduction of two plates together, rather than one descending beneath the other one, as is the case today.

Before 3000 Ma ago there were probably only a few small microcontinents, perhaps originating as volcanic island arcs. Most of these micro-continents were then captured by Late Archaean orogens to form larger continents, and this model is invoked to explain the large area of Precambrian crust encompassed in the North American Craton (Chapter 12). Computer simulations of the mantle at this time suggest that it may have had a strong thermal-stratification with the upper and lower mantle convecting heat independently. Subducted crust may have accumulated at the thermal boundary (660 km) between the upper and lower mantle. As the mantle cooled and this boundary became less pronounced, subducted crust may have periodically avalanched, as catastrophic events, towards the core, initiating a series of mantle plumes (super plumes). In fact, the catastrophic avalanching of subducted crustal slabs appears to have been a key feature of early supercontinental cycles, creating 'super' subduction zones attracting and accreting continents from afar. One particularly marked episode appears to have resulted in the creation of the Earth's first supercontinent around 2700 Ma ago in the Late Archaean. In a period of rapid crustal growth, numerous island arcs were created, many of which were trapped within and accreted onto the growing supercontinent, producing the characteristic relationship of granitic gneisses (the early crust) with greenstone belts (the arc volcanics and associated sediments). At least one more supercontinental cycle associated with mantle plumes and rapid crustal accretion occurred in the Early Proterozoic. By the Mid-Proterozoic a large supercontinent (Rodinia or

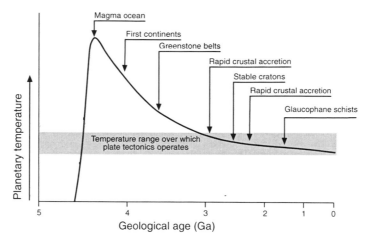

Figure 10.1   *Average planetary temperature through Earth history. [Modified from: Huggart (1997) Environmental Change, Routledge, Figure 3.12, p. 83]*

Palaeopangaea, Box 12.2) is thought to have formed, again associated with mantle plumes and episodes of rapid crustal growth.

As the mantle continued to cool, the thermal stratification within it decreased such that some time during the Mid-Proterozoic deeper subduction started, causing the mantle to convect more uniformly. The start of deep subduction is marked by the occurrence of high-grade glaucophane schist belts (Figure 10.1), as glaucophane is a metamorphic mineral usually associated with high pressure regimes. From this time on, conventional plate tectonics appears to have operated and the mantle convected heat more uniformly. This meant that the accumulation of cold crustal slabs at the junction of the upper and lower mantle was less marked and consequently catastrophic slab avalanches, and the resultant mantle plumes, appear to have decreased in importance relative to more conventional plate tectonics through the Phanerozoic. However, it must be emphasised that this is just one possible scenario, and only with further investigation will a clearer understanding emerge. What is clear, however, is that conventional plate tectonics was dominant from the Mid-Proterozoic onwards.

The distribution of the main cratons and continental units from the Late Proterozoic and through the Phanerozoic is relatively well understood, although a range of rival configurations have been proposed for the Late Proterozoic (Box 12.2). The changing pattern of continental blocks during the Phanerozoic is illustrated in Figure 10.2. This pattern of changing continents is perhaps best viewed as part of a supercontinental cycle in which the continents were initially widely dispersed at low latitudes in the Early Palaeozoic, before they coalesced in the Late Palaeozoic, through a series of orogenic episodes, to form a single supercontinent (Pangaea). As we have seen, this is probably the third such supercontinent to have existed, the other two forming and dispersing during the Precambrian. Pangaea was again fragmented in the Mesozoic to give the dispersed pattern of continents present today (Figure 10.2). Fortunately the geography of this last supercontinental cycle is

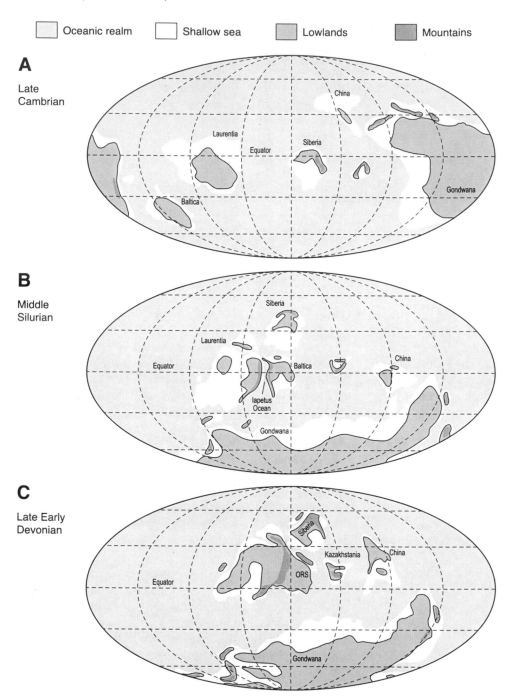

Figure 10.2 *The distribution of continental blocks through the Phanerozoic. It is important to note that areas of these continental blocks were flooded at various times. [Based on information in: Scotese (1991)* Palaeogeography, Palaeoclimatology, Palaeoecology **87**, *493–501; Scotese & McKerrow (1990)* Memoir of the Geological Society of London *No 12, 1–21]*

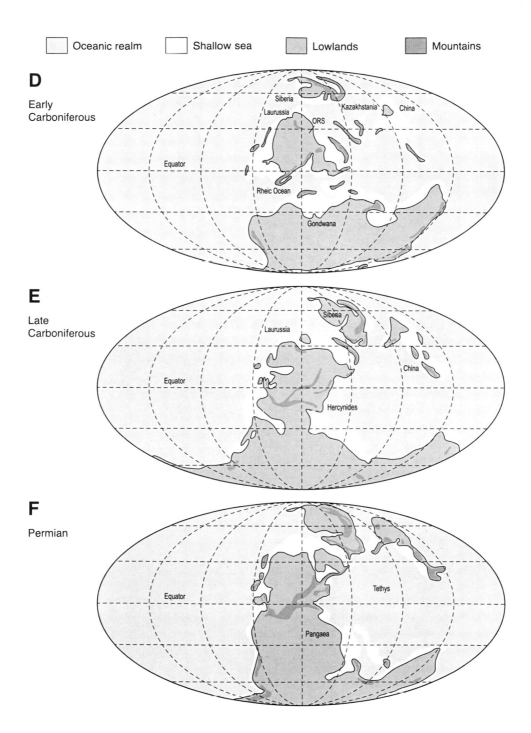

Figure 10.2 *Continued*

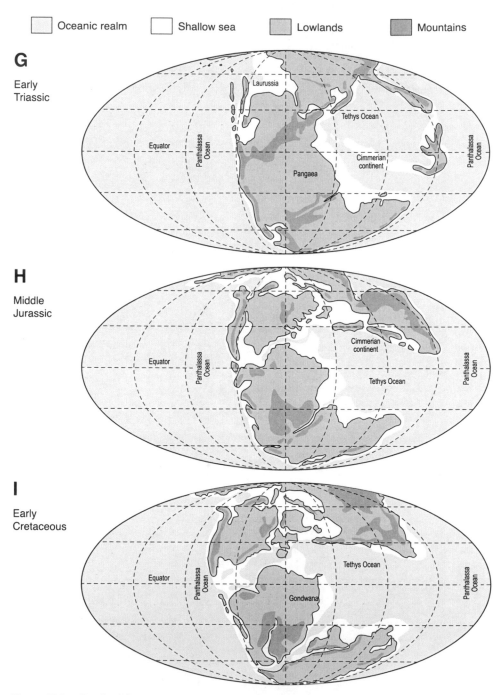

Figure 10.2   *Continued*

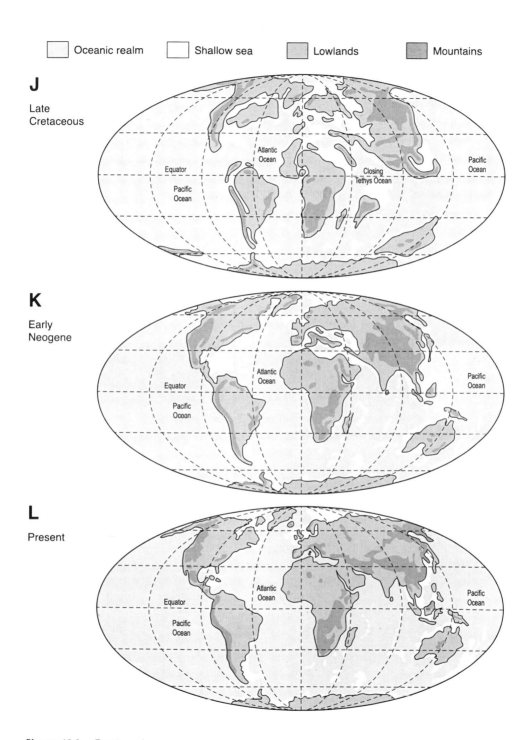

Oceanic realm   Shallow sea   Lowlands   Mountains

J Late Cretaceous

K Early Neogene

L Present

Figure 10.2   *Continued*

better documented and geologically constrained than its Precambrian counterparts, and we have a clear picture of its geological influence in the Phanerozoic.

The cause of supercontinental cycles is unclear. Some authorities have suggested that they may result from the contrast in the thermal properties of continental and ocean crust. Continental crust is only about one-half as efficient at conducting heat as oceanic crust, so that if a supercontinent such as Pangaea covers a significant part of the Earth's surface, then heat will build up in the mantle below it. Given time the build up of heat beneath the supercontinent may lead to uplift, rifting and continental dispersal. As ocean basins form during continental dispersal, mature subduction zones will develop and basin contraction may start. This will lead to a transition from continental dispersal to continental aggradation and the development of another supercontinent. The problem with this type of model is that the thermal contrast between ocean and continental crust is, according to recent modelling, unlikely to be sufficient to cause supercontinents to rift. Moreover it is not clear what causes the continents to converge to form a single supercontinent. In view of this, more recently the supercontinental cycle has been linked to plume tectonics.

It has been widely noted that there is an overlap between the dispersal and assembly phase of supercontinents; as one supercontinent is being dispersed, another is being assembled. In part this has been explained by the fact that mantle upwelling, a prerequisite for continental rifting and dispersal, causes thermal up-doming of the Earth's surface to form a geoid high. The geoid is the overall shape of the Earth itself, which is actually quite variable. As individual continents rift from supercontinents they will tend to move towards geoid lows, or towards cool, and therefore lower, areas of the Earth's surface. As continents converge in these geoid lows a concentration of subduction zones is likely. This will further cool the mantle as cold slabs of ocean crust descend. If these slabs become concentrated at the base of the upper mantle along the 660 km discontinuity then the process of catastrophic avalanching already described may occur, thereby generating mantle plumes (Figures 6.6 and 10.3). A likely result of this down-welling and avalanching is the pulling of plates towards subduction zones at accelerated rates, attracting and concentrating continents into a large supercontinent. In this type of model mantle plumes are created as a consequence of slab avalanches during the construction of the supercontinent and explain the episodic creation of supercontinents, particularly during the Archaean and Proterozoic when slab avalanching may have been more pronounced.

Once created, the supercontinent may act to shield the mantle from cooling by subduction, which is concentrated around the margins of the supercontinent. As the mantle beneath the supercontinent begins to warm a mantle plume may form, causing the supercontinent to rift and thereby restarting the cycle. Slab avalanching need not therefore be implicated in continental dispersal, although other variations of this model suggest that subduction, and slab avalanching around the margins of a supercontinent) may actually generate the mantle plume that ultimately leads to its demise (Figure 10.3).

In summary, the evidence suggests that the world's continents have experienced successive periods of continental assembly and dispersal associated with the repeated formation of a series of supercontinents, of which Pangaea is the latest and best

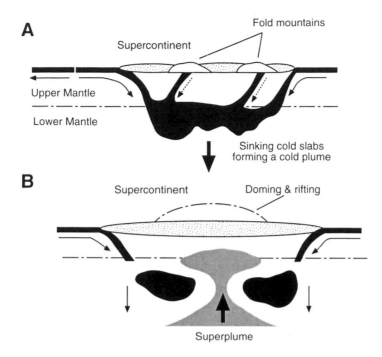

Figure 10.3   *The possible role of plume tectonics within the supercontinental cycle. [Modified from: Huggart (1997)* Environmental Change, Routledge, *Figure 3.15, p. 89]*

known. If the record of Pangaea is anything to go by, then it is clear that these giants have had a defining role in the development of the Earth's environmental systems.

10.2   Climate

The Earth's climate has varied throughout geological time. This is clear from the evidence provided by the stratigraphical record discussed in Section 5.3. However, the mechanisms by which climate has changed are poorly understood and numerous terrestrial as well as extra-terrestrial factors have been proposed to explain the Earth's climatic record. These explanations include variation in: topography, sea level, continental distribution, atmospheric carbon dioxide, solar radiation and the position of the Earth's solar system within the galaxy. Some of these factors are discussed below.

10.2.1   Climate through Geological Time

Using the palaeoclimatic tools outlined in Section 5.3 it is possible to obtain an impression of the climate which existed during a particular geological time interval. This is done by plotting the distribution of palaeoclimatically sensitive lithologies or fossils of a similar age on a palaeogeographical map of the period in question. When this information is combined with other sources of evidence, such as chemical isotope data, a reconstruction of the palaeoclimate can be obtained. A broad picture of the Earth's palaeoclimate over the last 500 Ma has been established in this way, but our knowledge of Precambrian climates is poor, since palaeoclimatic reconstruction becomes difficult with increasing age.

During the Late Proterozoic and Phanerozoic the Earth's climate appears to have oscillated between two stable states: one of global warming, the greenhouse state, and one of global cooling, the icehouse state. During the periods of global refrigeration, or 'Ice Ages', ice sheets dominated the high-latitude areas of the globe and left a clear record of their presence in the form of deposits and landforms. During greenhouse periods the Earth was usually ice free and there was generally a period of global warmth. The number of oscillations between these two states detectable in the stratigraphical record depends largely on the scale or detail with which the record is viewed.

On the broadest scale it is possible to recognise seven icehouse—greenhouse oscillations in the Phanerozoic, and we review these, and some of the evidence to support them, below. In addition, despite the fact that current knowledge of Precambrian palaeoclimate is poor, there is good evidence to suggest that the Earth was in an icehouse state—the so-called 'snowball Earth'—during the Proterozoic. This glaciation appears to have been one of the most extensive and long-lived periods of glaciation in the Earth's history (Box 12.3) and forms an event from which to begin a brief review of the Earth's palaeoclimatic history.

1. Icehouse World: Late Proterozoic (800–570 Ma). This period of glaciation is one of the most extensive within the geological record and may have lasted for up to 230 Ma. It is particularly puzzling because evidence of glaciation is not only found in rocks which formed at high latitudes, but also in those which formed at low latitudes of less that 45°. This period appears, therefore, to have been a particularly intense episode of global cooling. What is not clear, however, is the degree to which this represents a global event or simply a diachronous record of multiple glacial episodes in different parts of the globe.

2. Greenhouse World: Early Cambrian to Late Ordovician (570–458 Ma). During the Early Cambrian the Earth began to warm. The process of global warming appears to have been slow and reached a peak about 468 Ma ago. Low latitude temperatures at this time have been judged to be about 40°C, with the climate a little wetter than at present.

3. Icehouse World: Late Ordovician to Early Silurian (458–428 Ma). The first glaciers appear to have formed just after 458 Ma ago and developed into a full ice sheet glaciation around 10 Ma later in northern Africa, which was then situated close to the South Pole. Local, smaller ice centres developed in other parts of Africa and South America as the continent of Gondwana moved over the South Pole. The average low latitude temperature has been judged to be about 22°C, which is not too dissimilar from the present day average of 24°C. Other parts of the Earth appear to have experienced arid conditions during this period.

4. Greenhouse World: Late Silurian to Early Carboniferous (428–333 Ma). This Greenhouse period started with slow gradual warming which persisted until about 367 Ma ago. This period of warming is associated with an irregular increase in aridity which is illustrated by an abundance of evaporites and an absence of the humid indicators, coal and bauxite. There is also a pronounced increase in the amount of carbonates which are increasingly found at high palaeolatitudes (Figure 10.4).

5. Icehouse World: Early Carboniferous to Late Permian (333–258 Ma). This Late Palaeozoic cool mode was associated with the formation of high-latitude glaciers on many parts of the continent of Gondwana, which was located in the southern Hemisphere. This glacial episode may have lasted for 100 Ma and therefore represents the longest period of glaciation during the Phanerozoic. On this basis it is worthy of a fuller discussion here. This period of glaciation in the Southern Hemisphere is coincident with what has been described as 'the greatest episode of coal deposition in Earth history'. This depositional episode is recorded in the equatorial region of Laurussia (largely in North America and northern Europe), while Gondwana, much farther to the south, was being glaciated.

The glacial record of Gondwana is contained in intracratonic basins, the geometries of which are controlled by older structures formed during the construction of the

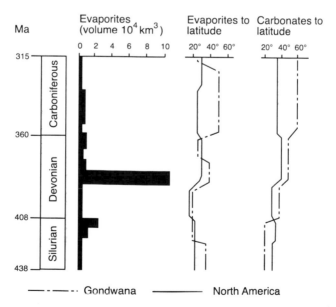

Figure 10.4 *Variation in climatic indicators during the Silurian to Early Carboniferous Greenhouse mode [Modified from: Frakes et al (1992) Climate Modes of the Phanerozoic. Cambridge University Press, Figure 4.3, p. 30]*

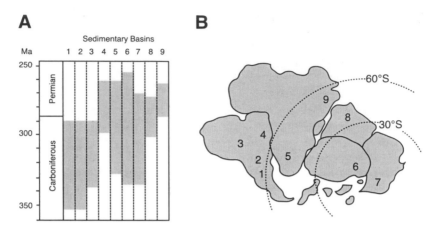

Figure 10.5   *The evidence for Late Palaeozoic glaciation.* **A:** *Correlation chart showing the age relationship of glacial strata in different glacial basins.* **B:** *Location of key glacial basins containing Late Palaeozoic glacial sediments. [Modified from: Eyles (1993)* Earth Science Reviews **35**, *Figures 16.1, 16.2 & 16.3, pp. 126, 127 & 128]*

continent in the Late Proterozoic (Figure 10.5). Many of the basins follow the lines of suture between the component cratons and were linked to the sea by epicontinental seaways. Reactivation of these ancient structures during the Carboniferous and Permian caused basin subsidence and preservation of glacial and glacially influenced marine strata. The stratigraphical record of this glacial episode is strongly diachronous, with glacial strata younging towards the east (Figure 10.5). This has been attributed to a gradual shift in the location of the main ice centres as Gondwana moved over the South Pole. Alternatively, it has been argued that this diachroneity may in part reflect not a climatic cause but a tectonic one, since preservation of the glacial strata is a function of basin subsidence which was not necessarily synchronous. In fact there is strong evidence to suggest that in some areas glacial erosion of upland areas may have taken place over a substantial interval before any glacial strata was actually preserved. A consequence of this is that the geological record of the glacial episode may in part be a record of basin subsidence and preservation, which may not have been coincident with the initiation of glaciation, rather than an accurate record of a climatic event. Cyclic fluctuations in the size of the Gondwana ice sheets have been implicated in driving a range of eustatic sea level changes during the Carboniferous, which is characterised by a variety of cyclic sedimentary systems (cyclothems) preserved in the coal facies of Laurussia. However, in most cases these cyclic systems can be explained by local or regional tectonics, and it is unlikely that the Carboniferous ice sheets had a significant role to play in such cycles which would require large sea level shifts.

6. Greenhouse World: Late Permian to Early Cenozoic (258–55 Ma). This has been generally regarded as an extensive period of global warmth when conditions were considerably more equable than at present. However, recent work has suggested that within this long period of warmth a cool mode can be identified, although not of the severity of previous episodes. One can therefore subdivide this long period into three shorter ones. First, on the basis of facies and fossil data for the interval between the Late Permian and Middle Jurassic (258–187 Ma ago), there was a distinct warm mode. The influence of the large continental landmass of Pangaea may have been important at this time, with extensive areas of this huge landmass in which rainfall was very low and temperatures very high.

Second, between the Middle Jurassic and middle part of the Cretaceous (187–105 Ma ago) there was a period of cooler, and more seasonal, conditions. The recent discovery of large dropstones, postulated to have been ice-rafted, in strata of this age may indicate that seasonal ice may have existed during this period at high latitudes (Box 10.1), but as yet no direct evidence of glaciation during this period has been recorded.

Finally, following this cooler period there was a return to a warm humid climate between the middle part of the Cretaceous and the Early Cenozoic (105–55 Ma ago). This period was one of the warmest of the Phanerozoic with the average global temperature probably about 6°C higher than it is today. Temperatures were high enough for the poles to be completely ice free and home to both forests and large

---

Box 10.1
Dropstones in the Cretaceous

The Mesozoic is traditionally considered to have been one of the longest Greenhouse periods within the Phanerozoic, with global warmth and ice free poles. Recently, the work of Frakes and Francis (1988) have challenged this model. The Lower Cretaceous Bulldog Shales of the intra-cratonic Eromanga Basin of central Australia are laminated marine mudrocks which contain large rock clasts of up to 3 metres diameter. The question raised by these clasts is this. Laminated mudstones are commonly produced in still waters, while transport of such outsized clasts requires very swift currents. The mudrocks do not contain any evidence of strong current activity, so what is the agent of transport? Frakes and Francis suggest that the clasts must have been dropped into the mud from a raft, and conclude that the most likely rafting agent is that of seasonal sea ice, which picks up clasts in the winter and releases them in the summer. This has implications for the supposedly warm climate of the Cretaceous, which according to these conclusions must have had at least seasonal ice in central Australia during the Cretaceous (palaeolatitude 65° and 78°S). This nicely illustrates the power of small facies interpretations in determining global climatic patterns. There is, however, one word of caution. In his voyage on H.M.S. *Beagle*, Darwin observed that 'the inhabitants of the Radack archipelago, a group of lagoon-like islands in the midst of the Pacific, obtained stones for sharpening their instruments by searching the roots of trees which are cast upon the beach (Darwin 1879, p. 461). This word of caution has been explored by Bennett and Doyle (1996), who have challenged the idea of Cretaceous cooling on the basis of dropstone evidence alone. Bennett et al (1996) review the wide range of mechanisms by which dropstones may be produced, many of them non-glacial in origin.

Sources: Frakes, L.A. and Francis, J.E. 1988. A guide to Phanerozoic cold polar climates from high-latitude ice-rafting in the Cretaceous. *Nature* **333**, 547–549. Darwin, C. 1879. *Naturalists' Voyage Round the World*. (Second Edition). John Murray, London. Bennett, M.R. and Doyle, 1996. Cretaceous cooling: fact or wishful thinking? *Terra Nova* **8**, 182–185. Bennett, M.R., Doyle, P. and Mather, A.E. 1996. Dropstones: their formation and significance. *Palaeogeography, Palaeoclimatology, Palaeoecology* **121**, 331–339.

reptiles (in the Cretaceous). The mid-Cretaceous was also a time of globally high sea levels and extensive areas of shallow shelf seas, ideal for the development of moderate climates, due to the increased evaporation and precipitation. Finally, after a sea level maximum at the close of the Cretaceous, climate appears to have begun to cool, with increased seasonality.

7. Icehouse World: Early Cenozoic to Present (55–0 Ma). The first stages of the Cenozoic cool mode started with global cooling in the Early Cenozoic (Eocene) some 55 Ma ago (Figure 10.6). From this time the Earth's climate gradually cooled from the warm conditions of the Mesozoic to the cool glacial climates of today. This cooling continued throughout the Cenozoic and polar ice caps developed in Antarctica and Greenland. This period was followed, 2 Ma ago, by the growth of mid-latitude ice sheets in both Scandinavia and North America. Most of the British Isles was glaciated and much of our current landscape was formed during this period. These mid-latitude ice sheets have accumulated and decayed in a cyclic fashion throughout the last 2 Ma. Periods of ice expansion are known as Glacials, while periods in which the mid-latitudes were ice free are known as Interglacials. The Earth is currently within an Interglacial and even with current predictions of anthropogenic climate change, a Glacial interval, along with mid-latitude ice sheets, is likely to return within the next few thousand years. This cyclic expansion and contraction of mid-latitude ice sheets during the Quaternary is due to

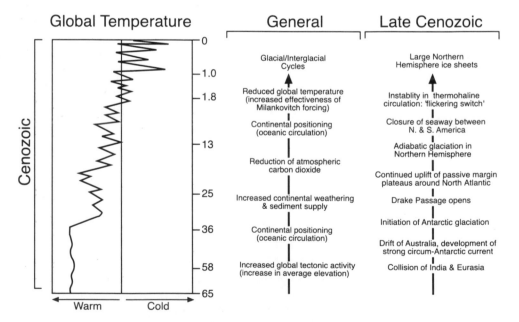

Figure 10.6   *Cenozoic cooling and the tectonic factors responsible. [Modified from:* **A:** *Boulton (1993) In: Duff (Ed.)* Holmes' Principles of Physical Geology, *Chapman & Hall, Figure 21.18, p. 446;* **B:** *Eyles (1993)* Earth Science Reviews **35**, *Table 21.1, p. 185]*

systematic variations in solar radiation caused by changes in the Earth's orbit, known as Milankovitch radiation variations. A detailed record of the growth of the polar ice caps during the Cenozoic and of the cyclic expansion and contraction of the mid-latitude ice sheets during the Quaternary is recorded within the oxygen isotope record (Box 5.4).

One can identify several similarities between each icehouse and greenhouse state within the stratigraphical record. Icehouse states appear to start with a long period of cooling which culminates in the growth of high-latitude ice sheets. Such periods are normally terminated by a sudden episode of climatic warming. They are invariably associated with the presence of extensive continental landmasses at high-latitudes and the growth of polar ice sheets. The formation of mid-latitude ice sheets may be relatively rare and confined to the Quaternary and Late Proterozoic 'Ice Ages'. The variation in intensity of Icehouse periods is hard to explain, but the presence of mid-latitude ice sheets in the Quaternary 'Ice Age' may be a function of Cenozoic mountain building. In contrast, greenhouse conditions appear to start rapidly and are normally terminated in a more gradual fashion via a period of cooling which takes place at different rates in different geographical locations. Greenhouse conditions are loosely correlated with periods of geological time when: (1) the atmosphere possessed a higher concentration of greenhouse gases, such as carbon dioxide; (2) sea-level was high and therefore large areas of the continents were flooded; and (3) the Earth's ocean basins were either very large or well connected.

In summary, the Earth's climate appears to have oscillated between periods of global cooling (icehouse) and global warming (greenhouse). It is important, however, to note that superimposed on this large scale oscillation are smaller scale temperature fluctuations within a given warm or cool phase. This scale of variation is well illustrated by the growth and decay of mid-latitude ice sheets during the Quaternary. It is important to note that the existence of an Icehouse period does not necessarily imply that all of the Earth was glaciated, but simply that the climate was cooler and that the equatorial belt may have been more restricted.

The change from one climatic state to another must have been associated with enormous changes in the climate system, while the relative homogeneity of each mode, persisting over tens of millions of years, indicates a large degree of inertia to change within both climatic systems. What causes this large scale oscillation in climate? One can identify four broad types of climatic forcing which may be responsible: (1) extra-terrestrial causes; (2) variation in the amount and distribution of solar radiation received; (3) variation in the composition of the atmosphere; and (4) the nature of the Earth's surface—the distribution of land and sea.

1. Extra-terrestrial causes. The impact of meteorites or 'bolides' has been put forward as a possible explanation for the change between Icehouse and Greenhouse states. The impact of a large meteorite would cause global cooling through the production and injection of masses of dust into the upper atmosphere, blocking off the sun's radiation. The impact of a large extra-terrestrial body in an ocean might also promote higher temperatures through the vaporisation of large amounts of water. The increased cloud cover which would result from this would insulate the Earth, preventing heat loss and thereby causing a global warming. Although such ideas are currently fashionable, for the most part they remain untested. There is increasing evidence that a meteorite impact was responsible for the mass extinction at the end of the Mesozoic, the so-called K–T boundary, and that one of the consequences of this impact was a short period of severe climate similar to that postulated in the 1980s for a 'nuclear winter' following a nuclear war. However, there is no convincing evidence at present to suggest that meteorite impacts are able to switch the Earth's climatic mode or produce sustained change in the Earth's climate system.

2. Radiation variations. A more plausible explanation for climate change involves variation in the amount of solar radiation received in the upper part of the atmosphere. Variation in the input of solar radiation to the Earth occurs at two broad scales: (1) long frequency variation associated with galactic cycles, which may drive icehouse–greenhouse oscillations; and (2) short frequency variations associated with rhythmic changes in the Earth's orbit, which may drive climate change within both icehouse and greenhouse modes.

It has been suggested that the oscillation between Icehouse and Greenhouse modes may be a function of long frequency, but high magnitude variations in the amount of solar radiation received by the Earth. The solar system rotates around the centre of the galaxy once every 300 Ma, a period known as the galactic year, and in so doing it passes through two stationary 'clouds' of hydrogen rich particles which may cause fluctuation in the amount of solar radiation received by the Earth. Consequently, every 150 Ma the Earth may receive less solar radiation. It has been suggested that the record of 'Ice Ages', or icehouse, states has an approximate periodicity of 150 Ma. This is an interesting hypothesis, but at present the geological record of glaciation and palaeoclimate is not precise enough to allow any firm conclusions to be made.

High frequency variations in insolation are due to rhythmic variations or 'wobbles' in the Earth's orbit, and may drive climate change within both icehouse and greenhouse modes. These high frequency orbital oscillations are known as Milankovitch radiation variations after the astronomer who first calculated them in 1941, although their existence had been postulated much earlier by Croll. The solar radiation, and its seasonal variation, received by the Earth varies in a cyclic fashion with rhythmic changes in the Earth's orbit (Figure 10.7). Three orbital cycles can be identified: (1) eccentricity of the orbit; (2) axial tilt (obliquity); and (3) precession of the equinoxes. Eccentricity of the orbit involves changes in the Earth's orbit from an elliptical orbit to a circular one every 95 000 and 400 000 years. Today the Earth is 146 million km from the sun on 3 January (perihelion), and 156 million km on 4 July each year (aphelion). This results in a 6 per cent difference in the amount of solar radiation received between the two seasons. When the orbit is at its most elliptical this figure rises to 30 per cent. The more elliptical the shape of the Earth's orbit becomes the more exaggerated the seasons become in one hemisphere and the more moderated in the other. Axial tilt involves changes in the inclination of the Earth's axis of rotation. Today the axis is inclined at an angle of 23.5°, but varies every 40 000 years between 24.5° and 21.5° (Figure 10.7B). This again has the effect of exaggerating the seasons, particularly at high latitudes; high angles of tilt increase the seasonal contrast. Finally, the precession of the equinoxes or solstices involves a number of interacting cycles. On a frequency of 27 000 years the Earth's axis of rotation moves around a full circle, due to the gravitational pull of the Sun and Moon, consequently the time of the year at which the Earth is nearest to the sun changes through time (Figure 10.7C). On a frequency of 105 000 years precession of the Earth's orbit occurs. Interaction of the two precession cycles gives a combined precession cycle of approximately 23 000 years, while the interaction of precession and eccentricity gives a small peak at 19 000 years.

The three main orbital cycles—eccentricity, obliquity and precession—interact, sometimes combining sometimes cancelling each other out, to cause subtle variations in the intensity and most importantly the seasonality of incoming solar radiation through time. It is the influence of Milankovitch cycles on the seasons which is of

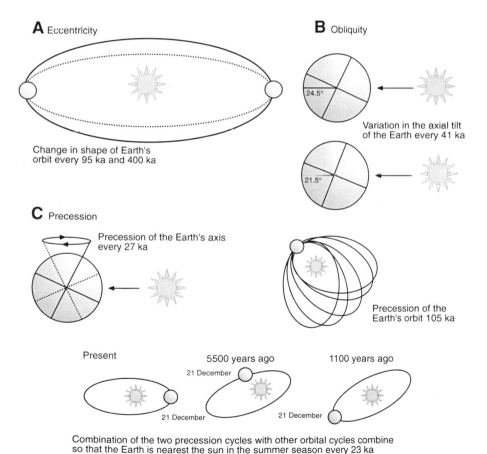

Figure 10.7 *Cyclic changes in the Earth's orbit cause, so-called Milankovitch radiation variations, variations in the amount of solar radiation received by the Earth through time. Three main cycles can be identified: eccentricity, obliquity and precession. See text for detailed discussion*

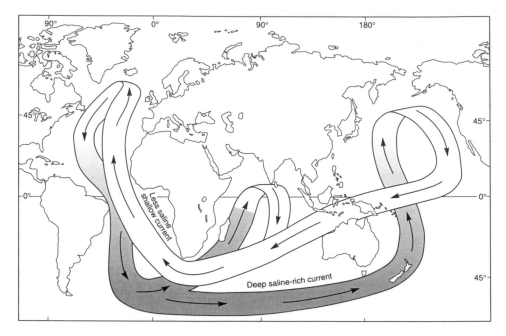

Figure 10.8   *The thermohaline circulation in today's oceans. [Reproduced with permission from: Lowe & Walker (1997)* Reconstucting Quaternary Environments. *Longman, Figure 7.27, p. 363]*

greatest importance. For example, in the context of glaciation, reduced summer insolation will facilitate the survival of winter snow, leading to the growth of ice sheets. Another consequence of the influence on seasonality of Milankovitch radiation variations is that the greatest impact occurs in the northern Hemisphere at latitudes of between 60° and 65° N. This just happens to also coincide, during the Cenozoic, with the latitude with the greatest percentage of land area relative to ocean. Seasonal contrasts are most marked on land, since ocean water has a much greater thermal inertia.

Milankovitch radiation variations have driven the pulse beat of glaciation, causing ice sheets to wax and wane, deserts to spread and contract, throughout the later part of the Cenozoic (Figure 10.6). This link was established through the spectral analysis of the oxygen isotopes recorded in ocean sediments which give a proxy record of global ice volume (Box 5.4). The rhythms in the oxygen isotope record correspond to the variation in solar radiation predicted by the three orbital cycles. Although this link is well established and accepted there is still a problem; how are relatively weak insolation changes, concentrated at latitudes of 60° to 65° N, responsible for shifts in global climate? Moreover the interplay of glacial and interglacial episodes during the Cenozoic 'Ice Age' is asymmetric with rapid termination of glacial periods, which are not a feature of the Milankovitch radiation cycles. Consequently, some form of amplification mechanism is required.

A range of different amplification mechanisms have been suggested, and the impact of mountain building (Box 10.2) and the uplift of rifted passive margins around the

---

Box 10.2
Why do mid-latitude ice sheets form in some 'Ice Ages' but not in others?

Certain 'Ice Ages' in the geological record appear to have been more intense than others. For example, the Cenozoic icehouse, and possibly the Carboniferous and Late Proterozoic Icehouses, were associated with mid-latitude ice sheets, while other episodes of Icehouse conditions experienced only polar glaciation. It has been suggested that the intensity of an 'Ice Age' is determined by the pattern of atmospheric circulation. The Cenozoic Icehouse is believed to have been intensified by the modification of atmospheric circulation through mountain building during the Cenozoic. Within the Northern Hemisphere there are two regions of extensive highland: (1) the Tibetan Plateau and the Himalayan mountains of Asia and (2) the Western Cordilleras of North America. Both these areas are broad, plateau-like bulges on which narrower mountain ranges are superimposed and they only reached their present elevation within the last million years. The large-scale circulation of the atmosphere is strongly influenced by these areas of uplift. They produce large scale standing waves (Rosby waves) within the upper atmosphere of the Northern Hemisphere. These waves occur in the lee of the western edge of the North American Cordilleras, to the east of the mountains of the European continent, and in the lee of the Tibetan plateau. Without these mountains the pattern of atmospheric circulation would be much smoother. Consequently the Late Cenozoic uplift of these mountainous areas produced a series of waves and troughs within the pattern of atmospheric circulation of the Northern Hemisphere. There is therefore a strong north–south component to the circulation, which during a period of Milankovitch cooling would tend to draw down polar air into mid-latitudes and thereby intensify the cooling episode in these areas. The occurrence and distribution of mountains within an 'Ice Age' may therefore have important consequences in determining its intensity.

Source: Ruddiman, W.F. and Kutzbach, J.E. 1990. Late Cenozoic plateau uplift and climatic change. *Transactions of the Royal Society of Edinburgh: Earth Sciences* **81**, 301–314.

North Atlantic are key factors in global cooling during the Cenozoic (Figure 10.6). However, attention has recently focused on the pattern of deep ocean circulation, driven primarily by contrasts in salinity between the Atlantic and Pacific oceans, in amplifying orbital radiation rhythms. This so called thermohaline circulation results from the distribution of land, sea and mountains around the two great ocean basins of today, which is ultimately a function of plate tectonics. The contrast in salinity between the Atlantic and the Pacific is also in part due to the distribution of mountains relative to prevailing winds around the basins. In the Atlantic this distribution is such that more evaporation than precipitation occurs. In part this is due to the Gulf of Mexico which acts like a large simmering saucepan producing warm saline water. This, coupled with the cooling of water beneath the sea ice of the northern Atlantic and off Antarctica, drives a pattern of sub-subsurface water circulation (Figure 10.8).

The key feature of this thermohaline circulation is the Gulf Stream, or North Atlantic Drift, which moves warm water northward in the North Atlantic region and maintains a more equable climate. During glacial periods the thermohaline circulation appears to break down. Milankovich cooling of the North Atlantic margins, aided by the tectonic uplift of these passive margins, may have initiated a small ice cover, influenced continental run-off, and increased the distribution of sea ice. All of these factors have influenced the salinity of the North Atlantic and may have caused periodic collapse of the thermohaline circulation, thereby leading to further cooling. A positive feedback loop of this sort is just the type of amplification mechanism required. In addition, changes in ocean circulation will affect the supply of nutrients and therefore the biological productivity of the oceans, this may in turn cause variations in the level of atmospheric carbon dioxide, again modulating climate change. The thermohaline circulation has been described as a 'flickering switch', driven by Milankovitch radiation variations, and in turn driving global climate change. It has also been implicated in very rapid high frequency climate changes at sub-Milankovitch scales. A complex web of feedbacks and processes operate to amplify orbital radiation variations into episodes of global climate change and the relative importance of each component is still uncertain and subject to debate.

Although Milankovitch radiation variations have driven the pulse beat of climate change during the Cenozoic, they are not in themselves responsible for the shift between icehouse and greenhouse conditions identified in the geological record. The long-term stability of Milankovitch rhythms is also open to question. Mathematical analysis suggests that the rhythms can be calculated accurately for perhaps 100 Ma, before chaotic elements in the system begin to complicate them. As such, the degree of error involved in extrapolating Milankovitch cycles into the past puts into question the validity of this approach, and is a question of some debate. It is also worth noting that Milankovich forcing during the Cenozoic appears to have been most effective at a latitude of 60°N; the impact of this was probably amplified by the Cenozoic concentration of continental surfaces at this latitude. Given different palaeogeographies, the latidudinal effectiveness of Milankovitch radiation variation may have been very different either amplifying or reducing its effectiveness. Equally the latitudinal range over which orbital forcing is most marked may have varied in the past. Despite this, Milankovitch cycles have been recognised in the Mesozoic Greenhouse, driving variations in biological productivity which are then preserved as cycles in sedimentary record (Box 10.3).

3. Variations in Atmospheric Carbon Dioxide. The budget of atmospheric carbon dioxide may help determine the Earth's climatic state (Figure 10.9). There is a strong correlation, for example, between periods of relatively low levels of atmospheric carbon dioxide and glaciation, such as in the Late Proterozoic and Carboniferous, although the Ordovician glaciation is actually associated with relatively high levels of this gas. Carbon dioxide is a greenhouse gas and its proportion within the atmosphere will influence global temperatures: the more carbon dioxide, the greater the greenhouse effect and the warmer the climate. At the simplest level, atmospheric carbon dioxide concentrations are dependent on: (1) the rate of volcanic out-gassing, which is a function of plate tectonics and (2) changes in the rate of continental weathering. Sea floor spreading, and the associated volcanic activity, provides the main input of carbon dioxide (volcanic gas) to the atmosphere, while chemical weathering of silicate rocks removes or 'scrubs' carbon dioxide from the atmosphere. Carbon dioxide is removed by chemical weathering because rain dissolves the gas within the atmosphere to produce an acidic solution that breaks down silicate rocks and locks the carbon molecules into weathering products. Periods of continental rifting and rapid sea floor spreading may therefore be associated with warmer climates, while periods of orogenesis and consequent erosion, revealing fresh silicate rocks for weathering, will lower carbon dioxide content and lead to cooler climates.

The Cretaceous Greenhouse period has been correlated with high carbon dioxide values (Figure 10.9) produced by the rapid rate of continental rifting and sea floor spreading associated with the opening of the Atlantic Ocean. The high sea levels of this time would also have reduced the area of silicate rocks available for weathering.

Box 10.3
Milankovitch radiation rhythms in non-glacial strata

Sedimentary cycles are a common feature of the stratigraphical record and have been attributed to a wide range of processes. As the role of Milankovitch radiation variations in driving climate change during the Cenozoic 'Ice Age' was established, and the scepticism about orbital-forcing reduced, attention has turned to the potential of orbital-forcing as a mechanism with which to explain many of these high resolution sedimentary cycles. Examples of ancient sedimentary cyclicity attributed to orbital-forcing have now been described from the Precambrian and every period of the Phanerozoic in a wide range of different facies (Weedon 1993; Table below). In some cases this is linked to glacially induced fluctuations in sea level, but in others is a function of changes in biological productivity and continental run-off induced directly by orbital-forcing. The methodology involved in inferring an orbital forcing is dependent on the resolution of the chronostratigraphical record. Essentially a succession is dated and the number of sedimentary cycles in a given period of time determined and divided by the length of time; if the resulting frequency corresponds to one of the main Milankovitch frequencies an orbital-forcing is inferred. The significance in recognising an orbital forcing lies in the potential to use such cyclicity to correlate between different sedimentary successions, a process which is sometimes referred to as cyclostratigraphy. Orbital-forcing is global and consequently is an example of event stratigraphy.

Kauffman (1988) provides an example of a Milankovitch-driven facies model for the North American Cretaceous, which involves asymmetrical bedding couplets, or rhythms, with varying lithological composition depending upon the depositional environment involved. Strata representing cool wet phases of the cycle lie at the base of each couplet, while lithologies representing warm dry phases lie at the top as illustrated in the diagram below. Kauffman (1988) proposed three tests to prove their origin as Milankovitch-driven events: (1) they must have durations and stratigraphical repetition compatible with the periodicity of orbital cycles; (2) they must be consistent with the likely climatic impacts of orbital-forcing; and (3) they must be regional or global events present in a range of different facies. Using these three measures Kauffman (1988) suggests that there is strong evidence to support the presence of an orbital forcing within the Cretaceous of the Western Interior Basin of North America.

Examples of models which link global climatic variation to sedimentary cycles

Facies	Link between climatic and sedimentary cycles
Evaporitic	Aridity/humidity variations control evaporite accumulation
Lacustrine	Runoff and humidity control lake depth and hence facies
Clastic shore	Position of shoreline related to glacioeustatic sea level
Deltaic	Position of delta front related to glacioeustatic sea level
Platform carbonate	Exposure, submergence and growth related to glacioeustatic sea level
Shelf carbonate	Facies related to water depth and glacioeustatic sea level
Clastic shelf	Grain size dependent on turbulence under control of storminess or sea level
Hemipelagic	Carbonate percentage, related to clastic input from runoff, percentage carbon related to productivity and/or water stratification
Pelagic	Carbonate percentage related to productivity and/or supply of clastics and/or sea floor dissolution
Turbiditic	Clastic supply related to sea level in relation to shelf or carbonate platform

Box 10.3
Continued

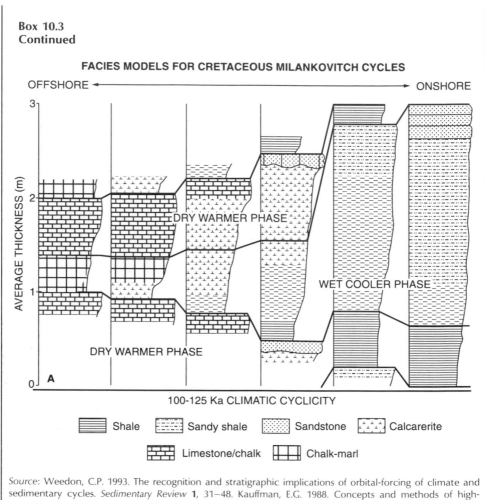

FACIES MODELS FOR CRETACEOUS MILANKOVITCH CYCLES

OFFSHORE ◄─────────────────────────────► ONSHORE

100–125 Ka CLIMATIC CYCLICITY

Shale   Sandy shale   Sandstone   Calcarerite

Limestone/chalk   Chalk-marl

Source: Weedon, C.P. 1993. The recognition and stratigraphic implications of orbital-forcing of climate and sedimentary cycles. *Sedimentary Review* **1**, 31–48. Kauffman, E.G. 1988. Concepts and methods of high-resolution event stratigraphy. *Annual Review of Earth and Planetary Science* **16**, 605–654. [Table modified from: Weedon (1993) *Sedimentary Review* **1**, Table 3.2, p. 36; Diagram modified from: Kauffman (1988) *Annual Review of Earth and Planetary Science* **16**, Figure 15, p. 641.]

Land plants control rates of silicate weathering, since humus produced by their decay is acidic, accelerating weathering and may therefore indirectly control levels of atmospheric carbon dioxide. Possible invasion of Proterozoic landmasses by blue-green algae, and the radiations of early land plants in the Silurian and Devonian, and of drought adapted grasses in the Eocene have been charged with facilitating glaciations in the Late Proterozoic, Late Palaeozoic and Cenozoic, respectively. However, in practice, the system which controls the level of carbon dioxide within the atmosphere is more complex than is suggested in this simple treatment.

The latitudinal position and distribution of continents may also affect carbon dioxide levels. Recent computer simulations comparing 'cap-worlds' in which all the continents are concentrated at the poles, with 'ring-worlds' where they are

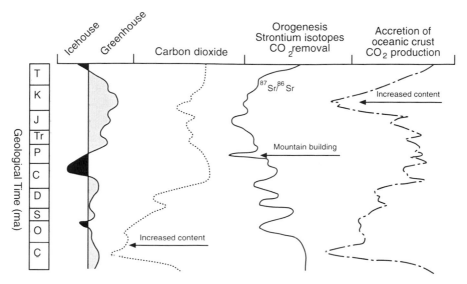

Figure 10.9 *Correlation between atmospheric carbon dioxide levels, the strontium isotope record, and the pattern of icehouse–greenhouse climatic oscillations. [Based on information from: Eyles (1993) Earth Science Reviews* **35**, *Figure 21.1, p. 182]*

aggregated at the equator, are instructive here. Chemical weathering is enhanced at the equator, due to the warmth and availability of moisture relative to the poles. Consequently, a 'ring-world' would be more effective in 'scrubbing' carbon dioxide, and according to the computer models would be 10° to 25°C cooler than the 'cap-world' scenario. Postulated Late Proterozoic continental configurations actually resemble a 'ring-world' and with an average continental latitude of just 21° and an average global temperature would be just 17°C. As we have seen, the Late Proterozoic was associated with an extensive episode of glaciation.

What emerges from this discussion is that a link may exist between cooler global climates and enhanced weathering, often associated with orogenesis. The correlation between orogenesis and icehouse conditions can be tested using the strontium isotope record (Box 4.4 and Figure 10.9). It has been confirmed that high levels of erosion and fluvial input into the world's oceans are generally associated with higher $^{87}Sr/^{86}Sr$ values, while lower values tend to occur with increased hydrothermal activity at spreading centres. As Figure 10.9 shows, there is some broad agreement between icehouse modes and peaks in the $^{87}Sr/^{86}Sr$ record, demonstrating the possible link between increased weathering and cooler climates, although as with any generalisation there are a number of exceptions. The role of tectonic activity in lowering atmospheric carbon dioxide is relatively well understood for the Cenozoic Icehouse interval, and may provide a key to understanding the role of atmospheric carbon dioxide in older glaciations (Figure 10.6). In particular, the elevation of the Tibetan Plateau over the last 40 Ma has had a dramatic impact on the flux of weathered debris and consequently on the reduction of carbon dioxide levels.

The atmospheric budget of carbon dioxide provides, therefore, an important factor in climatic forcing and may help to explain some of the variation between icehouse and greenhouse states recorded in the geological record. It is worth emphasising, however, that a range of other variables may also have had a role, particularly biological productivity. Productivity both on land and in the oceans is ultimately controlled by climate, distribution of habitats and ocean circulation and has in turn the potential to both draw-down and add to carbon dioxide levels in the atmosphere, thereby regulating, damping or amplifying trends in climate change. However, it is also important to note that variations in carbon dioxide cannot explain all shifts in climate mode, as the Late Ordovician glaciation well illustrates, occurring at a time when carbon dioxide levels were thought to be 16 times those of today.

4. The Distribution of Land and Sea. The nature of the Earth's surface may have had an important control on climate, through such factors as the latitudinal distribution of continents, the geometry of ocean basins, and the distribution of mountains. We have already seen how Cenozoic cooling has been linked to mountain building, both through its impact on carbon dioxide levels and its influence on atmospheric circulation (Box 10.2). In addition uplift can also be a direct cause of mountain glaciation (adiabatic glaciation). The availability of rifted passive margins as sites for glaciation around the margins of the North Atlantic uplifts during the Tertiary was probably a contributary, if not critical, factor in the initiation of Northern Hemisphere glaciation during the Cenozoic (Figure 10.6). We have also seen how deep ocean circulation during the Cenozoic may have played a key role in amplifying the effects of climate change. As we have seen, thermohaline circulation is a function of the distribution of mountains around the ocean basins of the Atlantic and Pacific as well as their latitudinal geometry. The pattern of near-surface circulation we see today in the oceans, has also evolved in the last 160 Ma since the break up of the supercontinent Pangaea (Figure 10.10). Since Australia broke free from Antarctica, opening the Southern Ocean, the Antarctic continent has become increasingly isolated from equatorial waters by the Antarctic Circumpolar Current, a process which was completed when the Antarctic Peninsula broke from South America approximately 25 Ma ago. Equally important to the pattern of circulation in the Atlantic was the closure of the gap between North and South America around 3 Ma ago.

Continental movement in the past will also have had a profound effect on ocean circulation and is ultimately a function of plate tectonics. Most greenhouse periods are associated with large well-connected oceans, allowing the poleward transport of heat. For example, the Cretaceous Greenhouse world was one of the warmest periods in the Earth's history, and occurred at a time when ocean basins were well connected and the resulting ocean circulation pattern may have allowed efficient heat transfer towards the poles. In addition, high sea levels and extensive continental flooding appear to be loosely correlated with periods of greenhouse conditions. Oceans and continental seas can absorb much more heat than land and also evaporate to give increased cloud cover, which helps retain heat. The high sea levels of the Cretaceous Greenhouse phase may also, therefore, be of considerable importance in determining the pronounced warmth of this period.

There have been a number of attempts to model the impact of different continental distributions on climate. One exercise has employed a general circulation model to

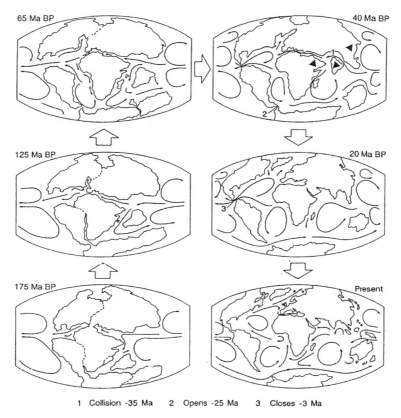

1 Collision -35 Ma   2 Opens -25 Ma   3 Closes -3 Ma

Figure 10.10 *Schematic reconstruction of the pattern of near surface ocean currents during the Mesozoic and Cenozoic [Modified from: Williams et al (1993)* Quaternary Environments, *Edward Arnold, Figure 2.2, p. 18]*

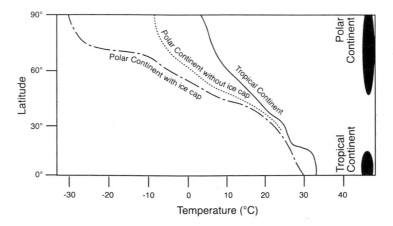

Figure 10.11 *A plot of average annual air temperature against latitude for three different continental distributions as predicted by a general circulation model. [Modified from: Barron (1992) In: Brown et al (Eds)* Understanding the Earth, *Cambridge University Press, Figure 24, p. 502]*

examine the global climate produced by both a large polar continent, with and without an ice cap, and a tropical continent. The contrast between a tropical and polar continental distribution was equivalent to 12°C, and this contrast was even more marked if a polar ice cap was present (Figure 10.11). The presence of high-latitude land provides a surface for the accumulation of snow. In turn the accumulation of snow may accelerate cooling, by increasing its albedo, and therefore the amount of solar radiation reflected, which reduces that left to warm the land surface. Further, high-latitude land may block the poleward penetration of warm ocean currents and therefore limit the poleward transport of heat. It is worth noting that these models do not incorporate the impact of continental distribution on carbon dioxide levels. We have already seen how polar continents may be associated with lower rates of chemical weathering and higher carbon dioxide levels, illustrating the complexity of the climate system.

The quest for an explanation of large scale climatic oscillations is not easy, and often complex. It is clear that it is unlikely that there is a single causal mechanism with which to explain all events and changes in the climatic mode of our planet, which tends to be associated with a range of factors. This is well illustrated by the Cenozoic Icehouse which appears to have resulted from a combination of elements: orogenesis, uplift of passive margins, as well as the occurrence of sensitive and potentially unstable patterns of ocean circulation. In turn, gradual cooling induced a state in which global climate could be driven by subtle orbitally induced insolation variations. The complex pattern of the Cenozoic Icehouse therefore provides a model of how earlier changes in global climate mode may have resulted from the interplay of a range of different factors. Perhaps what is clear is that the underlying, first order, control on climate is that of plate tectonics.

10.3 Sea Level

Sea level has varied throughout geological time and is an important control on the type of sedimentary facies deposited. Not only does it determine the extent of marine sedimentation but also the depth of water and therefore the nature of the sediments (e.g. beach sands or offshore muds) deposited within a basin. The mechanisms by which global (eustatic) sea level has varied are in many cases poorly understood, but the broad patterns of sea level variation appear to reflect changes in plate tectonic activity.

10.3.1 Sea Level Variation through Geological Time

Using the tools of facies analysis outlined in Section 5.1.2 one can obtain a broad picture of global sea level at a given point in time and space. By correlating the sea level records from different locations it is possible to recognise eustatic sea level events. These must be distinguished from local fluctuations in relative sea level caused by minor tectonic adjustments which alter land elevation. An estimate of global sea level variation throughout the last 500 Ma has been established (Figure 5.11), but our understanding of Precambrian sea level is poor, since the sedimentary record from this period is fragmentary.

The Vail sea level curve in Figure 5.11, constructed in the manner discussed in Section 5.1.2, shows both first order and second order cycles of sea level variation. The first order trend in sea level shows periods of high sea level during the Early Palaeozoic and the Cretaceous, separated by a period of much lower sea level. During the Cretaceous sea level may have been as much as 350 metres above its present level. Superimposed on this first order pattern is a second order sequence of major transgressions and regressions. About 14 second order cycles can be identified, each of which lasted between 10 and 80 Ma. During each of these second order cycles the sea level rises, slows to a standstill and then drops rapidly. Superimposed on this pattern of second order variation are numerous small scale sea level fluctuations which probably reflect local, rather than eustatic, variation. These local variations may be caused by fluctuations in the size of ice sheets during an 'Ice Age', or by the build-up of flexural stresses within moving plates which causes local tectonic adjustments which, in turn causes variation in land elevation.

Why has sea level risen and fallen episodically throughout the Phanerozoic? Eustatic sea level changes can be brought about by either: (1) changing the volume of ocean water, or (2) changing the volume of the ocean basins.

1. Changing the Volume of Ocean Water. This can be achieved in three ways. First the volume of ocean water can be reduced by increasing the volume of land ice. In fact, sea level can fall by up to 150 metres simply through the growth of a large ice sheet. This is very important in explaining the sea level oscillations experienced over the last 2 Ma of the Cenozoic Icehouse. Second, the desiccation of shallow marine basins may reduce sea level. For example, during the Neogene the Mediterranean was an enclosed sea which was subject to periodic desiccation with consequential sea level changes of up to 15 metres. The final mechanism by which water volume can be changed is through temperature variation. For every degree Celsius that the temperature varies, the sea level can change by up to 1 metre, due to the thermal expansion or contraction of the water mass.

2. Changes in Ocean Basin Volume. There are three ways in which the volume of the ocean basins may change. First, the mid-ocean ridges fill a significant part of the volume of an ocean basin. Sea floor spreading takes place at mid-ocean ridges, and variation in the rate of spreading can alter the volume of the ridge and therefore affect sea level. As new oceanic crust is formed and moves away from the ridge crest it cools, contracts and subsides. The new crust is hot and thermally buoyant and thus increases the topography and volume of the ridge. If the rate of sea floor spreading is rapid then new ocean crust is formed over a much larger area, and this means that the volume of the ridge is much greater than at ridges where sea floor spreading is slower. If the rate of sea floor spreading slowed or decreased then the size of the ocean ridge would decrease, with the result that the volume of the ocean basin would increase, leading to a sea level fall (Figure 10.12). Conversely, an increased spreading rate would lead to a sea level rise. In this way,

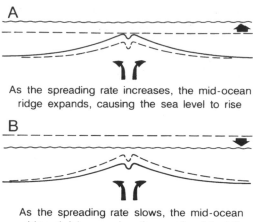

Figure 10.12   *The relationship of the rate of sea floor spreading to sea level fluctuations. [Modified from:* Lemon (1990) Principles of Stratigraphy. Merrill Publishing Company, *Figure 12.5, p. 306]*

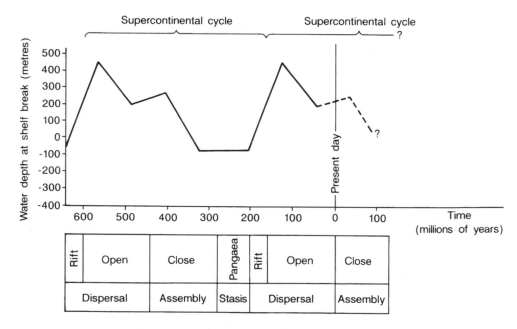

Figure 10.13   *The possible relationship between long-term eustatic sea level fluctuations and the supercontinental cycle. At times of continental assembly the average age of the ocean floor is at a maximum, spreading is slow, heat flow is low and therefore the ocean basins are deeper and eustatic sea levels lower. As a supercontinent splits up the average age of the ocean floor decreases, heat flow increases, spreading rates are high and as a consequence mean ocean depth decreases, causing high sea levels. [Modified from:* Worsley et al (1984) Marine Geology **58**, *Figure 13, p. 391]*

variations in the rate of sea floor spreading may cause global sea level to fluctuate by as much as 300 metres. Sea floor spreading rates do not appear to be uniform and their variation may account for many of the marine transgressions and regressions present within the stratigraphical record.

A second mechanism leading to volume changes may be that of continental rifting and collision. During rifting, continental crust is stretched and continental area increased, a condition which may create marine transgressive episodes. Conversely, subduction along continental margins, or continental collision, decreases continental area and leads to marine regression. It has been suggested that the first order pattern of sea level variation may be viewed in terms of the supercontinental cycle (Figure 10.13). The periods of high sea level during the Early Palaeozoic and the Cenozoic reflect periods of continental dispersion when sea floor spreading and continental rifting dominated, while the low sea levels of the Early Mesozoic reflect the formation of the supercontinent of Pangaea (Figure 10.13).

Finally, a third possible explanation for ocean volume changes is that of erosion, and more importantly, the resulting pattern of sedimentation. Increased sedimentation leading to the construction of large-scale features such as deltas could effectively displace the water mass of a basin and lead to a sea level rise. In practice, however, it is unlikely that this would have anything other than a regional effect in altering sea level.

Despite the range of possible explanations for eustatic sea level changes the pattern of first and second order variation recorded in Figure 5.11 cannot be completely accounted for. There is no single explanation for all sea level variations and it is possible that different mechanisms may have operated at different times.

10.4   Biosphere

The biosphere encompasses all parts of the Earth in which life is present. In essence, it is equivalent to the sum total of life on Earth (the biota) at any given time. From the first appearance of life on Earth, some 3500 Ma ago, the individual components of the biosphere have evolved and the nature of the biosphere has changed. Thus, the strata of different geological intervals each contain different assemblages of fossil animals (fauna) and of plants (flora). Fossils give an insight into the nature of the biosphere through geological time and in some cases, are significant rock forming materials themselves (Figure 10.14). The following section outlines the broad evolutionary processes that have shaped the Earth's biosphere and then examines the broad pattern of change through geological time.

10.4.1   Evolution and the fossil record

Evolution is the process by which species undergo gradual transformation through time. Although much of the modern study of evolution is tied to understanding the genetic processes of inheritance and species variation, it is only geology that can record the changes in evolving lineages (lines of descent of evolving organisms)

Figure 10.14   *Coccolith limestone from the Upper Jurassic of southern England. Average coccolith diameter 5 μm. [Reproduced with permission from: Young & Bown (1991)* Palaeontology **34**, *plate 1, p. 1]*

through time. Geology is particularly important in helping us to determine the broadest scale of evolutionary changes in the development of the biosphere.

The changes in organisms up to and including the creation of species are encompassed within microevolution. It is the biological study of genetics that enables us to understand the mechanisms of microevolutionary change. The development of an individual organism is ultimately controlled by its genes. Genes are composed of DNA and are normally located in the cell nucleus arranged along elongate, thread-like bodies known as chromosomes. The DNA molecule carries a chemical code known as the genetic code, which stores all of the information necessary for the function and development of an organism. As DNA is passed on from generation to generation, it provides the basis for the inheritance of characteristics and the continuity of life.

The gene pool is a collective name for all genes in a species or particular population. Evolution is expressed as a change in a population's gene pool through time. Natural selection is the process which operates on existing genetic variation (i.e. current species) to bring about successive changes in those species through time. Certain genes are lost from the gene pool because of the failure of individuals with those genes to reproduce and new genes are introduced through mutation. In this way natural selection takes place and the genetically determined characteristics that result from selection and which are in turn passed on through reproduction are known as adaptations.

New species arise from an existing species' gene pool when individuals form a population that diverges so much from other populations of the same species that the two groups can no longer interbreed: in other words when the interchange of genes between the two groups stops. This reproductive isolation is usually associated with geographical isolation of a population, or with an ecological restriction to a specific habitat. Put simply, if the individuals of one species are divided into two separate populations which become isolated from one another by a physical barrier then the flow of genes within the gene pool common to the two groups will be interrupted. In this case new species will eventually be created when individuals from the two groups have evolved so far that they can no longer interbreed. The creation of new species through such isolation is known to be extremely important in the development of new species.

In the past it was envisaged that evolution progresses gradually with many intermediate stages. This is a view which is very close to the ideas first expressed by Charles Darwin, the founder of modern evolutionary thought, in 1859. This model of evolution is often referred to as phyletic gradualism. In this model all the intermediate forms along an evolutionary path should be present; their absence in the geological record is assigned to the imperfection of the stratigraphical record. In recent years, however, scientists have questioned the traditional model of phyletic gradualism and have suggested that evolutionary change consists of periods of stasis, when no changes occur, punctuated by rapid periods of morphological development. Put another way this means that evolution is seen to occur in sudden bursts. This view of evolution has already been introduced in Section 4.1.3 (Box 4.3) and is known as punctuated equilibrium. The fossil records of many groups often display periods of stasis and rapid change consistent with the theory of punctuated equilibrium, although in some cases, rapid and gradual change may be seen to occur together within a single fossil group ('punctuated gradualism').

Species transitions of the sort described above are the domain of microevolution, but how do major new groups of animals and plants form? The geological record gives clear evidence of the diversification and development of new groups and evolution at this scale is often referred to as macroevolution. The origin of major groups and the direct relationship of micro- and macroevolutionary processes is still an area of active research, but it is clear that the summation of microevolutionary processes must lead to large scale changes in the biosphere. It is also clear that macroevolutionary processes may be a reflection of large scale environmental changes: changes that result in the operation of natural selection processes that change the gene frequency in a large number of organisms at the same time. Two macroevolutionary processes that appear to be of extreme importance in the development of the biosphere, are mass extinctions and adaptive radiations.

The fossil record provides a direct insight into the development of the biosphere through time. It is clear that there are periods when the standing diversity—the total number of different types of organism present on the Earth—has fallen dramatically over a short period of time. Species extinctions occur regularly in nature, but these mass extinctions involve numerous species and have had a dramatic effect on the evolution of the biosphere.

Mass extinctions appear to have occurred regularly throughout geological time. There have been at least five major mass extinctions in the history of life (Table 10.1) and several smaller ones (Figure 4.10). In fact, some authors have argued that mass extinctions may occur every 26 Ma and are therefore a periodic phenomenon associated with a specific, possibly extra-terrestrial, cause. This view has, however, received considerable criticism as the data set is derived from the compilation of published literature, and may have compounded errors. Quite apart from this criticism, it is clear that several of the peaks are out of phase with the postulated 26 Ma periodicity, and as such this hypothesis remains unproven (Figure 10.15). The cause of mass extinctions is uncertain and there are four main hypotheses currently available to explain them. What is clear from the record is that no one cause is responsible for all the known extinctions, although some form of climate change is a theme of many.

1. Sea Level Variations. As we saw in the previous section sea level fluctuates through time. The effects of sea level change are known to have had a profound

Table 10.1   *Major episodes of mass extinction in the Phanerozoic, together with candidate causes. Evidence for major meteorite impact is strongest only in the Late Cretaceous. [Percentage data from: Sepkoski (1982) Geological Society of America Special paper* **190**, *283–289]*

Mass extinction event	Percentage marine families lost	Candidate causes
End Cretaceous	15%	Meteorite impact Volcanic winter
End Triassic	20%	Climatic effects (warming) Sea level fall
End Permian	50%	Sea level fall Meteorite impact
Late Devonian (Frasnian- Famennian)	21%	Climatic effects (cooling) Meteorite impact
End Ordovician	22%	Sea level fall Climatic effects (cooling)

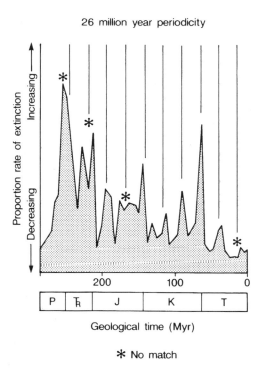

26 million year periodicity

∗ No match

Figure 10.15 *The postulated 26 Ma periodicity of extinctions. Note the mismatch of several extinction peaks with the periodicity. [Modified from: McGhee (1989). In: Allen & Briggs (Eds) Evolution and the Fossil Record, Belhaven Press, Figure 2.5, p. 36]*

effect on the abundance and diversity of the marine fossil record. For example, the sea level rise at the beginning of the Cambrian is thought to have influenced the radiation of marine organisms by increasing the size and availability of suitable habitats (ecospace), that of shallow marine shelf areas. Conversely, a fall in sea level may increase the competition for ecospace amongst marine organisms and therefore lead to species extinction. Such variations are thought to be largely responsible for the Ordovician, Permian and Triassic extinctions, although the causes of the sea level variations are themselves a subject for debate.

2. Climatic Change. The Earth's climate has fluctuated between icehouse and greenhouse states through geological time. Climate, and in particular temperature, is an important limiting factor on the distribution and diversity of organisms today. While it is possible to identify extinctions that are probably climate controlled, such as the extinction of large mammals in the Pleistocene through the onset of the 'Ice Age', it is perhaps more difficult to relate the mass extinction of both land and marine organisms directly to climate. What is clear, however, is that climate change may well have been the end result of periods of vulcanicity, or even the aftermath of extra-terrestrial impact, and as such, has a role to play in at least the end-Cretaceous extinction. Eustatic sea level rise or fall may also be a result of climate mode shift, with consequent effects on the biosphere.

3. Vulcanicity. A major volcanic episode may have caused the end-Cretaceous mass extinction, and this is still a major debating point in academic circles. The catastrophic eruption of volcanoes can lead to the input of large quantities of volcanic gases and ash into the upper atmosphere. This could produce a 'volcanic winter', a phenomenon which could be produced when ash particles and other aerosols are injected into the upper atmosphere. This would reduce the amount of solar radiation received by the Earth's surface, and in turn cause intense climatic cooling for a period of years. The end-Cretaceous extinction can be correlated with the eruption of large amounts of basaltic lava and it has been argued that these eruptions may have caused a short-term deterioration in the world's climate and thereby caused a failure of the global ecosystem. Such large scale outpourings of flood basalts may have occurred more than once in Earth history, and have been linked to mantle plume activity. This would suggest periodic outpourings in the more distant geological past, creating equivalent environmental changes which could force extinctions. Research continues in this area, but as yet, there is little evidence of flood basalt forced extinctions beyond the end-Cretaceous event.

4. Extra-terrestrial impact. The mass extinction at the Cretaceous–Palaeogene boundary (usually referred to as the Cretaceous–Tertiary or K–T boundary) is associated with a layer of clay rich in iridium (Figure 4.7). Iridium is commonly enriched only in meteorites and other extra-terrestrial bodies, and this has led to the suggestion that the mass extinction at the end of the Cretaceous may have been caused by the impact of an extra-terrestrial body. The impact of a meteorite would create what is known as an 'impact winter' similar to a 'volcanic winter' but induced by dust introduced into the atmosphere by the impact of a meteorite. It has also been argued that if the impact had taken place in an ocean, vaporisation of the water during collision would increase the global cloud cover leading to an accelerated greenhouse effect. Some authors have suggested that meteorite impacts occur at regular intervals as our solar system moves through the galaxy. These authors point to the apparent 26 Ma cyclicity of extinctions identified within the geological record as evidence of this, although as discussed most authorities have questioned this hypothesis (Figure 10.15).

The idea that an extra-terrestrial 'bolide' impact caused the end-Cretaceous extinction has received even greater attention since the discovery of the Chicxulub Crater on the Yucatán Peninsula in Mexico (Box 10.4). This buried crater is considered by many to represent the 'smoking gun' providing evidence of the impact of a large body capable of affecting global climate. Although most authorities accept its existence, debate still rages over its role relative to the flood basalt episode also apparent at the K–T boundary.

In actual fact, all four of these mechanisms for mass extinction may clearly be interrelated and it is unlikely that there is a single mechanism suitable to explain all. As an example of this it is conceivable that a climatic change (cooling) would lead to a sea

Box 10.4
Chicxulub Crater: the 'smoking gun' at the K–T boundary

The idea that a large extra-terrestrial bolide of immense size could have impacted with the Earth and precipitated a mass extinction was first seriously postulated by Alvarez et al in 1980. This hypothesis suggested that the impact of such a body would effectively alter the climate system and lead to ecosystem failure, associated with the phenomenon of an 'impact winter' created by the injection of dust from the impact and resulting 'wildfires' that would rage across the planet. Alternatively, greenhouse conditions could have ensued if the impact was into an oceanic setting, derived from the vaporisation of sea waters. Supporting evidence included the presence of the iridium-enriched layer, thought to be derived from space, the presence of quartz grains with a shock induced cleavage, and the presence of glassy droplets known as microtektites, indicating melting from the impact. All that was missing was a crater, the 'smoking gun' that would point towards the perpetrator of this global crime. This was provided with the discovery in the early 1990s of probably the largest crater on Earth, the 180–200 km diameter Chicxulub Crater, a structure of K–T boundary age buried deep in the Yucatán Peninsula in Mexico (Hildebrand et al 1991). The presence of this large circular structure has been demonstrated by boreholes, and by the results of geophysical investigations such as the Bougier gravity survey illustrated. The crater is developed in carbonate platform deposits, and is associated with acidic volcanics. Shocked quartz grains are present, and derived from the continental crustal rocks they could have been deeply sampled and widely distributed. A thick ejecta breccia blanket is present, and is close to two other sites with thick ejecta layers, in north-eastern Mexico and Haiti. Finally the microtektites seen in the boundary layers seem to match the chemistry of the acidic volcanics and carbonates of the crater stratigraphy. The discovery of this crater has helped advance the concept of a K–T boundary impact event, which may have caused mass extinction through enhanced greenhouse conditions. Although the debate still rages, the powerful evidence of the 'smoking gun' at the scene of the crime is compelling.

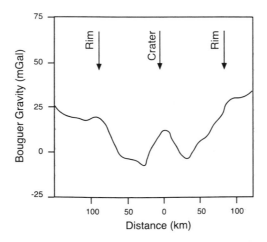

Sources: Alvarez, L.W., Alvarez, W., Asaro, F. and Michel, H.V. 1980. Extraterrestrial cause for the Cretaceous/Tertiary extinction. *Science* **208**, 1095–1108; Hildebrand, A.R. et al 1991. Chicxulub Crater: a possible Cretaceous/Tertiary boundary impact crater on the Yucatán Peninsula, Mexico. *Geology* **19**, 867–871. [Diagram modified from: Hildebrand et al (1991), *Geology* **19**, Figures 1, 3 & 4, pp. 867–869.]

level change (fall) through the growth of ice sheets. Any resulting mass extinction could therefore be caused by a combination of both processes. Although the mechanism of mass extinction is still hotly debated, and is an area of popular interest, it is clear that mass extinctions have occurred and have had a profound effect on the evolution of the biosphere.

Periods of mass extinction are often followed and thereby balanced by periods of evolutionary expansion which effectively 'restock' the depleted biosphere. These expansions occur with the development of a successful group of organisms which proliferates and expands its geographical range rapidly, over a short space of geological time. The success of these organisms is usually associated with the development of a new adaptation which gives them an advantage in re-populating the environmental space vacated by the organisms killed off during the mass extinction. In other cases radiations may reflect other changes in the environment. For example, stromatolites only became common in the Proterozoic as oxygen increased (Figures 10.16 and 10.17). In some cases this may also be triggered by the utilisation of an old adaptation in a new way. The resulting evolutionary expansions are usually referred to as adaptive radiations.

Together, mass extinctions and adaptive radiations effectively control the stocking and restocking of the biosphere through time. This pattern of extinction and radiation is illustrated in Figure 10.17B, which shows the clades or clusters of evolving groups

Figure 10.16   *Photograph of Late Proterozoic stromatolites from Orkney, Scotland. Layers of sediment are trapped by algal mats, which give the irregular laminations and structures shown in the photograph [Photograph: BGS]*

of animals and plants. These clusters represent the diversity of life on Earth at any given time, and they reflect the painstaking work of many specialists. Extinctions are indicated by rapid reductions in diversity (i.e. the width) while radiations are shown by similar rapid increases in the faunal diversity over a short interval of time (Figure 10.17B).

The Cambrian, Palaeozoic and Mesozoic–Cenozoic faunas represent the major stocks of marine invertebrates. These faunas are often referred to as evolutionary faunas, and represent a similar pattern of diversification among a variety of different organisms. The Cambrian Fauna, the first shelly biota, was replaced after the Ordovician mass extinction by the Palaeozoic Fauna, dominated by brachiopod-rich assemblages, which in turn was replaced after the Permian mass extinction by the Mesozoic–Cenozoic Fauna, dominated by molluscs (Figure 10.17B). The evolution of vertebrates and of plant life are represented by clades in Figure 10.17B, and both show patterns of extinction and radiation. An example of an extinction event and its aftermath can be illustrated by the extinction of the reptiles and the radiation of mammals at the end of the Cretaceous (Figure 10.17B). It is important to note, however, that adaptive radiations need not only be associated with extinctions. Particular examples of such radiations include the fish in the Devonian, and angiosperms in the Cretaceous. Both radiations are the result of the evolutionary success of a key adaptation, the development of shelly hard parts in marine invertebrates in the Devonian and the reproductive system (flowers and enclosed seeds) found in the angiosperms (flowering plants) during the Cretaceous.

In the following section the record of change within the biosphere through geological time is discussed.

10.4.2   The Biosphere through Geological Time

A record of the evolving biosphere is only possible through the presence of fossils within the geological record. The detail with which the biosphere can be reconstructed is usually limited by the quality with which the flora and fauna of a particular period are preserved. Fortunately, however, rare glimpses that approximate to the full diversity of life at specific points are given by the deposits known as Lagerstätten, a concept already introduced in Section 5.4 (Box 5.6). A Lagerstätte is a palaeontological 'bonanza' in which the condition of preservation has often been so favourable as to preserve both skeletons/shells (hard parts) and soft tissue (soft parts) of organisms.

The development of the biosphere may be differentiated into two major phases: (1) the Precambrian, in which life first formed and diversified, and (2) the Phanerozoic in which rapid diversification of the biosphere took place creating the complexities of life that surround us today.

1. Life in the Precambrian. Two major phases can be identified in the development of life in the Precambrian: ancient life in the earliest or Archaean rocks and more complex life in the younger Proterozoic rocks.

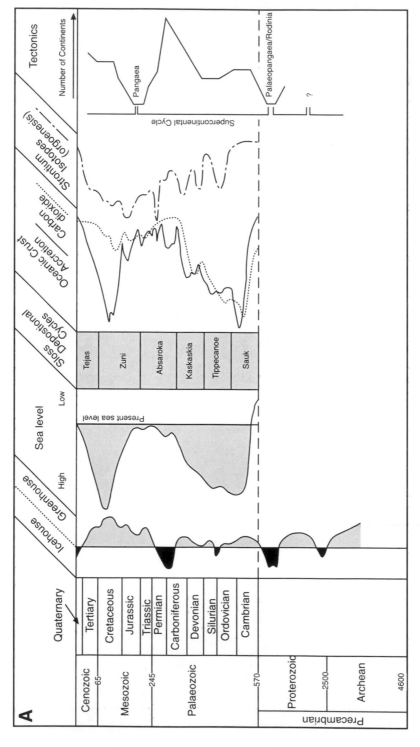

Figure 10.17 *Summary chart of the four key variables which determine the physical environment of the Earth.* **A:** *Climate, sea level and plate tectonics.* **B:** *The biosphere*

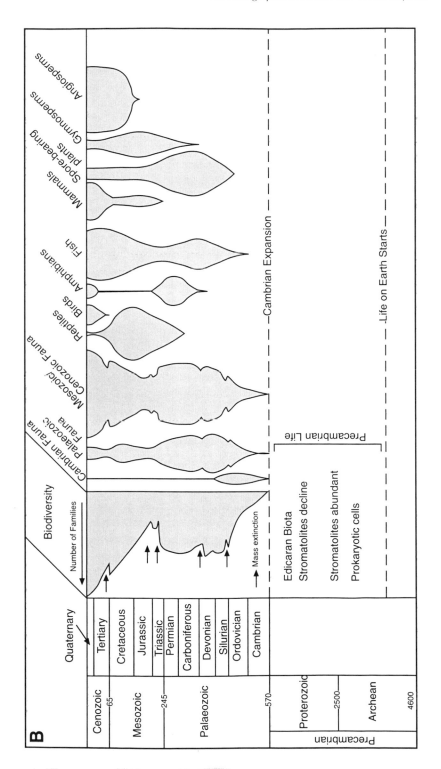

The first fossil organisms are recorded from rocks in Western Australia which have been dated at some 3500 Ma. These early cells, which are probably some form of bacteria, were able to survive in an anaerobic or oxygen-free environment, since it is thought that the Archaean atmosphere was composed of methane, carbon dioxide, and hydrogen sulphide produced by volcanic degassing as the Earth's crust cooled. The energy source of these early organisms is the subject of some debate. The conventional view is that they depended upon preformed organic molecules to serve as an energy source. An alternative is that these early organisms were dependent on an external energy source, but not the Sun. This idea has developed through the recognition of present-day bacteria associated with volcanic hydrothermal vents on the ocean floor. Here organisms meet their energy needs through oxidising hydrogen sulphide in solution in the hot waters surrounding the vents, and this process is known as chemosynthesis. Such bacteria, known as archaeobacteria, are now considered to be at the root of the tree of life, and their presence at vent sites gives strong evidence in favour of a hydrothermal origin for the first life. It is on this basis, and the fact that hydrothermal activity can be confidently predicted for Mars, that the search for fossil life on this distant planet is due to start in earnest.

In either hypothesis the evolutionary development of life is limited by resources, either by the amounts of preformed organic molecules, or by the distribution of suitable sites of inorganic energy. Therefore the appearance of a metabolism that utilised a truly renewable source of energy, such as light, was crucial to life on Earth. The development of photosynthesis which uses light as an energy source is one of the most significant events in the history of the biosphere. In photosynthesis solar energy is transformed to chemical energy when carbon dioxide is reduced by hydrogen, donated either by hydrogen sulphide in the case of certain anaerobic bacteria, or by water in the case of cyanobacteria and virtually all other photosynthetic organisms. Sulphur is released when hydrogen sulphide is used, while oxygen is released when water is the hydrogen donor. The first photosynthetic organisms lived in the early anaerobic atmosphere of the Earth. The development of photosynthesis is marked in the geological record by the occurrence of stromatolites (Figure 10.16). Stromatolites are formed by mats of cyanobacteria which trap sediment and thereby form mounds of sediment which are recorded in the sedimentary record. Their formation is closely associated with photosynthetic organisms. The development of a photosynthetic pathway which utilised water as the source of hydrogen and released oxygen as a by-product lead to a gradual, but radical, change in the Earth's atmosphere. By the end of the Archaean the Earth's atmosphere was being enriched in oxygen by photosynthetic organisms.

In the Proterozoic the Earth's atmosphere had become increasingly oxygen-rich as a product of the cumulative effect of photosynthesis. In many ways the subdivision of the Precambrian into the Archaean and the Proterozoic approximates to the change from an anaerobic (oxygen-poor) to aerobic (oxygen-rich) atmosphere. Probably the most important evolutionary step during the Proterozoic was the appearance of the first eukaryotic cell. This type of cell not only has a nucleus and complex of chromosomes, but also a means of sexual reproduction and therefore a method of genetic variation and of natural selection. It is not known exactly when this event occurred, but it could not have taken place in the anaerobic environment of the Archaean as eukaryotes require oxygen. From this type of cell multicellular organisms (metazoans) developed and these have been found in the geological record as far back as 1300 Ma ago. The first evidence of tracks and trails left by metazoans are to be found in rocks 900 Ma old. Following this in the late Proterozoic is a distinct assemblage of multicellular organisms, known as the Ediacara Biota, which was first found in the Ediacara Hills of Australia and dated at 600 Ma old. This biota and others like it (e.g. the Charnian Biota in England) are distinct from living organisms today, as although some members appear to resemble worms and jellyfish, others have a peculiar quilted structure which is unknown in animals today. This biota was extinguished at the end of the Proterozoic, apparently leaving no direct descendants.

2. Life in the Phanerozoic. The beginning of the Phanerozoic is marked by what is perhaps the most dramatic adaptive radiations, commonly referred to as the Cambrian expansion or explosion. This 'explosion' refers to the rapid diversification of animals with hard parts (i.e. shells and skeletons) and took place at the start of the Phanerozoic following the extinction of the Late Proterozoic Ediacara biota. This diversification may have been assisted by the effects of the widespread Late Proterozoic glaciation, which had two effects. First, it increased the mixing of water masses and therefore the supply of nutrients. Second, sea level increased significantly after the end of this 'Ice Age' and flooded large areas of continental margin creating shallow seas. The uptake of phosphate and then carbonate in these shallow seas by Cambrian organisms allowed them to form skeletal material for the first time.

In the early part of the Palaeozoic life was exclusively restricted to marine environments and was dominated by trilobites. Trilobites are related to arthropods, which are animals that have a segmented body, jointed limbs and are covered by a hard organic skeleton or shell (Figure 4.2). Crabs are typical arthropods. Trilobites dominated the marine environment in the Cambrian (the Cambrian Fauna), but at its end they suffered a mass extinction associated with a fall in eustatic sea level. Survivors of this event allowed a renewed radiation of the trilobites in the Ordovician, but for the rest of the Palaeozoic and particularly following a second, major extinction at the end of the Ordovician, shelly fossils, particularly brachiopods, dominated the marine environment (the Palaeozoic Fauna). Animals with backbones (vertebrates) appeared in the Late Cambrian in the form of primitive fish, which continued to diversify throughout the later part of the Palaeozoic. The land was first colonised by arthropods during the Ordovician. This became possible through the diversification of land plants which acquired the adaptive advantage of a rigid stem and tissue which was capable of carrying nutrients throughout the plant (vascular tissue). A famous Lagerstätte, the Scottish Rhynie Chert, gives insight into the form of these early plants and more significantly of the land-dwelling animals which they supported during the Devonian. Continued evolution of the woody and root tissue of plants allowed the development of coal swamps in damp humid locations. These early plants were tied to aquatic habitats, because their reproductive system required water. The first land dwelling vertebrates were amphibians derived from fish in the Early Carboniferous and were closely followed by the first reptiles.

Three major extinctions took place in the Palaeozoic: (1) at the end of the Ordovician, which decimated marine faunas, (2) at the end of the Devonian and (3) at the end of the Permian which effectively wiped out 50% of marine fauna (Table 10.1) and most of the early mammal-like reptilian fauna found on land. Each of these extinctions was matched by a subsequent adaptive radiation (Figure 10.17B).

The Mesozoic era started in the aftermath of the dramatic Permian extinction, coined by some as the day the Earth nearly died; it is estimated that 90% of marine species were killed-off. The nature of the shallow marine fauna changed: the Palaeozoic Fauna, dominated by brachiopods, was replaced by a fauna dominated by molluscs, the Mesozoic–Cenozoic Fauna, still characteristic of the shallow-marine environment today. Reef building corals, like those of today, appeared for the first time and replaced the niche left by the extinction of the Palaeozoic corals. Cephalopods, hunting molluscs such as squid, became an important predator in the world's oceans. On land, the gymnosperms (naked seed) plants radiated widely, through the development of a reproductive system that was not dependent on water. Consequently this allowed the colonisation of dry lands. The flowering plants, angiosperms (enclosed seed) appeared at the end of the Mesozoic era and are unusual because they resulted from an adaptive radiation which was not tied to a mass extinction. The angiosperms co-evolved with certain insects, which provided the means of fertilising the seed and consequently the development of the angiosperms is matched by the diversification of insect life. The vertebrates continued to diversify, most notably the dinosaurs and large reptiles dominated the land and sea. An extinction and subsequent radiation took place at the end of the Triassic, killing off much of the shallow-marine fauna, as well as many of the early reptiles. The best documented and most controversial mass extinction took place at the end of the Cretaceous and led to the demise of the dinosaurs, the marine reptiles and many species of marine molluscs and plankton.

The Cenozoic era followed this mass extinction and saw the development of the biosphere as we know it today. The most notable radiation, after the end-Cretaceous extinction, was that of the mammals which restocked the niches vacated by the reptiles. In much the same way the birds filled the gap left by the extinction of flying reptiles. The angiosperms were largely unaffected by the end-Cretaceous extinction and furnished the diversity of plant life we find today. The shallow-marine sea floor dwelling (benthic) fauna was largely unaffected by the end-Cretaceous extinction, but bony fish took over from the cephalopods as the dominant predator of the world's oceans. Many large mammals suffered extinctions towards the end of the Cenozoic which may reflect the onset of the Cenozoic 'Ice Age' or the first appearance of human hunters.

10.4.3   The Biosphere as a Control on the Stratigraphical Record

The Biosphere has contributed to the development of the Earth's stratigraphical record in two ways: first, through the modification of the local and even global environment and second, through the production of primary geological materials.

The development of much of the Earth's present atmosphere was probably a direct result of the summation of the by-products of photosynthesis. Today, the atmosphere could be restocked with oxygen in just a few thousand years, through photosynthesis. In the Precambrian the development of the atmosphere took some thousands of millions of years. An oxygen-rich atmosphere obviously sustains growth but also has an effect on the nature of the minerals and sediments that are deposited. Red beds are indicative of the oxidisation of iron minerals, and are common only from the Late Proterozoic onwards. Significantly, living organisms have an impact on the rate at which the land surface erodes and evolves. Prior to the colonisation of the land in the Early Palaeozoic there was little or nothing available to bind the surface of the soil and consequently the rates of erosion would have been high. With the evolution of land plants, rates of erosion and therefore of landscape change would have been reduced dramatically. Those areas where the vegetation cover was nearly complete would have experienced a relative fall in erosion and sediment production.

Living organisms also have a direct effect on the nature of sedimentation. Sea floor and land dwelling organisms disturb and mix accumulating sediment, while the bodies of such organisms enhance the rate of sedimentation at certain locations. In this way, the biosphere contributes directly to the stratigraphical record through the production of rocks of organic origin. The majority of carbonate rocks, for instance, are the product of the accumulation of vast numbers of shelly fossils, on either the microscopic or macroscopic scale (Figure 10.14). Chalk in particular is a carbonate rock of organic origin which has developed through the accumulation of countless millions of the microscopic skeletal remains of planktonic marine plants. This deposit is widespread across Europe and America and is a significant part of the stratigraphical development of these regions. Reefs and other carbonate accumulations are also an important feature of the stratigraphical record. This type of organic deposit illustrates at one level how the biosphere has had a direct input into the development of the stratigraphical record.

10.5 Summary of Key Points: the Earth through Time

We suggested at the start of this chapter that the stratigraphical record can be interpreted in terms of four variables: plate tectonics, climate, sea level and the biosphere. These four variables combine to determine the nature of the rocks deposited or formed at any point on the surface of the Earth. The stratigraphical record is therefore a record of these four variables, which have varied systematically throughout Earth history. In this chapter we have explored the rhythms of change present within each of these variables and obtained an impression of the way in which the Earth has evolved over the last 4600 Ma. There is much about this story that is incomplete or still unknown and our knowledge of the Precambrian is poor. It is also clear that these four strands of Earth history are interdependent and interact to a greater or lesser degree. For example, plate tectonics has an important influence on climate and sea level, which in turn has significantly contributed to the development of the biosphere, and feeding back, the biosphere has exerted its own influence in helping to moderate and evolve the Earth's climate (Figure 9.1).

The strands of this story of our planet have been brought together in Figure 10.17 in an attempt to summarise the key events of Earth history. This summation represents a key to what was happening to the Earth at any point in time and provides a model of the physical development of the Earth's crust, climate, sea level and biosphere. If one was to take a time slice across this summary chart then the rocks formed or being deposited across the globe at that time would be the product of the subtle interplay of the global plate tectonic setting, sea level, climate and the evolving biosphere. Our current understanding of Earth history has only been achieved through the efforts of numerous 'stratigraphical detectives' who have pieced together the evidence from across the world using the stratigraphical tool kit introduced in Part One.

10.6   Suggested Reading

Many of the subjects covered in this chapter are introduced and discussed in the very readable book by Van Andel (1985). The chapter by Fisher (1984) in *Catastrophes and Earth History* also provides a useful review of this subject. The detailed literature on the subjects reviewed in this chapter is vast and what follows is not intended to be a definitive guide, but aims simply to draw attention to some of the main sources of information.

Plate tectonics. Windley (1993) discusses the application of the plate tectonic model to the geological record with particular reference to the Precambrian. The distribution of the Earth's continents during the Palaeozoic is reported in Scotese and McKerrow (1990) and for the Mesozoic in Scotese (1991). More recent papers by Dalziel (1997) and Roger (1996) review more recent developments and controversies. The possible mechanisms underlying the supercontinental cycle are discussed at an introductory level by Nance et al (1988) and in detail by Gurnis (1988).

Palaeoclimate. Frakes (1979) provides a good review of palaeoclimates through geological time, a story which is updated in Frakes et al (1992). Eyles (1993) provides a superb review of the Earth's glacial record and of the mechanism responsible for it. The chapter in Duff (1993) on palaeoclimate provides a good introduction to some of the possible mechanisms of climate change, particularly in the context of the Cenozoic 'Ice Age'. Barron (1992) provides a more sophisticated discussion of this subject. The work of Hey et al (1990) gives an insight into the possible effects of continental distribution on the global climate, while the paper by Summerhayes (1990) introduces a series of papers in the *Journal of the Geological Society* on palaeoclimate (Volume **147**, pp. 315–392). The book by Imbrie and Imbrie (1979) is a good guide to Milankovitch radiation variations and their climatic impact. Wilson et al (2000) provides a good, accessible review of recent Cenozoic climate change.

Sea level. An accessible introduction to eustatic sea level variation in the geological record is provided by Lemon (1992). Hallam (1984) is a classic paper in this field and has much useful information, while Worsley et al (1984) discuss the possible linkage between the supercontinental cycle and eustatic sea level. Hallam's textbook provides a readable overview of the subject (Hallam 1992).

Biosphere. Stanley (1989) provides a good general account of the development of the biosphere from a geological perspective, while in contrast Bradbury (1991) provides a biological view of the evolving biosphere. Fortey (1997) provides a highly readable account of the development of life on Earth, while McAlester (1977) is a useful account, although a little dated. Bock and Goode (1996) is an interesting set of papers which examine the origins of life. Both Allen and Briggs (1989) and Donovan (1989) contain papers which examine the development of life through mass extinctions and adaptive radiation. Finally, Briggs and Crowther (1989) is still probably the most up-to-date source of information on the development of the biosphere through geological time.

References

Allen, K.C. and Briggs, D.E.G. 1989. *Evolution and the Fossil Record.* Belhaven Press, London.

Barron, E.J. 1992. Palaeoclimatology. In: Brown, G.C., Hawkesworth, C.J. and Wilson, R.C.L. (Eds) *Understanding the Earth.* Cambridge University Press, Cambridge, 485–505.

Bock, G.R. and Goode, J.A. (Eds) 1996. *Evolution of Hydrothermal Ecosystems on Earth (and Mars?).* John Wiley, Chichester.

Bradbury, I. 1989. *The Biosphere.* Belhaven Press, London.

Briggs, D.E.G. and Crowther, P. 1989. *Palaeobiology: a Synthesis.* Blackwell, Oxford.

Dalziel, I.W.D. 1997. Neoproterozoic–Paleozoic geography and tectonics: review, hypothesis, environmental speculation. *Geological Society of America Bulletin* **109**, 16–42.

Donovan, S.K. 1989. *Mass Extinctions: Processes and Evidence.* Belhaven Press, London.

Duff, P. McL.D. (Ed.) 1993. *Holmes' Principles of Physical Geology.* (Fourth Edition) Chapman & Hall, London.

Eyles, N. 1993. The Earth's glacial record and its tectonic setting. *Earth Science Reviews* **35**, 1–248.

Fisher, A.G. 1984. The two Phanerozoic supercycles. In: Berggren, W.A. and Van Couvering, J.A. (Eds) *Catastrophes and Earth History.* Princeton University Press, Princeton, 129–150.

Fortey, R. 1997. *Life: an Unauthorized Biography.* Harper Collins, London.

Frakes, L.A. 1979. *Climates throughout Geologic Time.* Elsevier, Amsterdam.

Frakes, L.A., Francis, J.E. and Syktus, J.I. 1992. *Climate Modes of the Phanerozoic.* Cambridge University Press, Cambridge.

Gurnis, M. 1988: Large-scale mantle convection and the aggregation and dispersal of supercontinents. *Nature* **332**, 695–699.

Hallam, A. 1984. Pre-Quaternary sea-level changes. *Annual Review of Earth and Planetary Sciences* **12**, 205–243.

Hallam, A. 1992. *Phanerozoic sea-level changes.* Columbia University Press, New York.

Hay, W.W., Barron, E.J. and Thompson, S.L. 1990. Global atmospheric circulation experiments on an Earth with polar and tropical continents. *Journal of the Geological Society of London.* **147**, 749–757.

Imbrie, J. and Imbrie, K.P. 1979. *Ice Ages: Solving the Mystery.* Harvard University Press, Cambridge, Massachusetts.

Lemon, R.R. 1990. *Principles of Stratigraphy.* Merrill Publishing Company, Columbus.

McAlester, A.L. 1977. *The History of Life.* (Second Edition) Prentice Hall, New Jersey.

Nance, R.D., Worsley, T.R., and Moody, J.B. 1988: The supercontinental cycle. *Scientific American* **259**, 72–79.

Rogers, J.I.W. 1996. A history of continents in the past three billion years. *Journal of Geology*, **104**, 91–107.

Scotese, C.R. 1991. Jurassic and Cretaceous plate tectonic reconstructions. *Palaeogeography, Palaeoclimatology, Palaeoecology* **87**, 493–501.

Scotese, C.R. and McKerrow, W.S. 1990. Revised world maps and introduction. Memoir of the Geological Society of London **12**, 1–21.

Stanley, S.M. 1989. *Earth and Life Through Time*. Second Edition. Freeman, New York.

Summerhayes, C.P. 1990. Introduction: palaeoclimates. *Journal of the Geological Society of London* **147**, 315–320.

Van Andel, T.H. 1985. *New Views on an Old Planet*. Cambridge University Press, Cambridge.

Wilson, R.C.L., Drury, S.A., and Chapman, J.L. 2000. *The Great Ice Age: Climate Change and Life*. Routledge, London.

Windley, B.F. 1993. Uniformitarianism today: plate tectonics is the key to the past. *Journal of the Geological Society of London* **150**, 7–19.

Worsley, T.R., Nance, D. and Moody, J.B. 1984. Global tectonics and eustacy for the past two billion years. *Marine Geology* **58**, 373–400.

11

The North Atlantic Region: a Journey Through Time

The aim of the next four chapters is to illustrate how plate tectonics, climate, sea level and biosphere have interacted through geological time to create the pattern of Earth history which is preserved by the stratigraphical record. We use one region of the Earth's crust as an example of how the pattern of change in these four key variables determines the nature of the Earth's physical environment through time. This could be attempted for any part of the Earth's crust and is already being applied, albeit remotely, to the distant planets such as Mars. We confine ourselves to our home planet here, and consider the North Atlantic region. We have chosen this region deliberately, but could easily have chosen any of the other great continents: Australasia, Asia, Antarctica, or South America. Each of these has a rich geological heritage, a heritage that can be teased apart in order to reveal the complex interaction of the four key variables of plate tectonics, climate, sea level and the biosphere.

In Chapters 12 to 14 we tell the story of the geological development of the present-day North Atlantic region, and specifically the continental areas of north-western Europe and North America. In particular we shall concentrate on that area now separated by the Atlantic Ocean, and chart its history during the last four billion years, showing how this story is the product of the interplay of plate tectonics, climate, sea level and the biosphere through time. We shall, however, only discuss the key events in the geological development of our chosen region as they appear to us. Here and there we will adapt our focus in order to zoom in and out of specific areas which are illustrative of important principles or processes. On this basis, these chapters are not intended to be a definitive treatise on the stratigraphy of the North

Atlantic region, but rather to serve as an overview of the key events and episodes which characterise its stratigraphical record and which illustrate how global processes have created and shaped the Earth. The text should be read with reference to the summary charts of the global record (Figure 10.17).

The subject of stratigraphy, like most other scientific subjects, is constantly developing as new ideas and theories are proposed, changing our perspective on Earth history. The story we give below is, therefore, a current view of the stratigraphy of the present North Atlantic region, a view which will no doubt become more precise as more evidence is obtained and as the interpretative framework of models and ideas is improved.

11.1 The North Atlantic Region

The North Atlantic region as we know it today has had a relatively brief existence, especially so since the Atlantic itself only came into existence some 160 Ma ago during the Late Mesozoic. In fact, there have been at least two great oceans in the region, the Iapetus – named after the father of Atlas in Greek mythology – and the Atlantic, and the fates and fortunes of these have shaped the geology of northern Europe and North America. Other oceans, such as the Rheic, the Tethys and both the ancient and modern Pacific, have also had a role to play, and these are also incorporated into our story.

What is clear from this is that plate tectonics is of critical importance in its development, and the fold mountain chains of Europe and North America are the physical scars of the turmoil that the Earth has undergone in its long history of continuous change. As we have seen in Chapter 6, the early geologists of Europe and North America were well aware of the distinction between the continental interiors and the linear belts of sediments, often strongly folded, which were welded against the mostly ancient, metamorphic cratons (Figure 6.1). These cratons, termed shields because of the shape of the Precambrian outcrop in North America on a map, were demonstrably older and formed a kind of substrate to which other, younger rocks are added in a kind of collage. This idea still holds today, although as we have seen, the level of sophistication in our understanding has increased ten-fold with the birth and acceptance of plate tectonic theory. We now accept that plate tectonics provides the backdrop to the evolution of any continental region, and as we will see, it has had a major effect on the development of our region. It provides the framework for the development of the stratigraphical record, and as a consequence, continental construction dominates any account of the early part of the development of both North America and northern Europe. In turn, the biosphere, climate and sea-level have operated to produce the sediments and to fill the sedimentary basins created by plate tectonics, the detailed characteristics of which are discussed throughout the text where appropriate. However, in determining the Earth's global record, it is important to realise that other factors may overprint the clarity of this record. For example, due to plate tectonics our region has gradually drifted northwards, from the Southern to the Northern hemispheres, through geological time (Figure 11.1). This northward drift has had an impact on the climate and biota of the region as it drifted

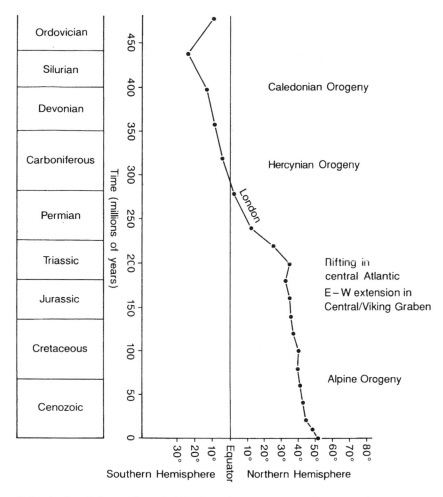

Figure 11.1 *A plot of the northward drift of southern Britain (London) since the Early Ordovician. [Modified from: Glennie (1990). In: Glennie (Ed.) Introduction to the Petroleum Geology of the North Sea, Blackwell, Figure 2.7, p. 43]*

through the tropics, which overprints the effect of the Earth's global pattern. It is for this reason that although the Carboniferous globe was in an icehouse state, northern Europe and North America was experiencing more balmy conditions, due to their equatorial location.

We have divided the story of the North Atlantic region into three blocks of time: (1) the Precambrian; (2) the Cambrian to Carboniferous, the period in which the supercontinent Pangaea was assembled and which encompasses both the Caledonian and Hercynian (Variscan) orogenic phases; and (3) the Permian to Recent, in which Pangaea was disassembled, the Atlantic opened and the mountain ranges of the European Alps, the Himalaya (Alpine Orogeny) and the Western Cordillera were formed. The first of these blocks of time is vast, over 4000 Ma, but in contrast to the other two we know very little about it, although the North

American craton preserves a record of the earliest evolution of the Earth's crust. The end of the Precambrian represents a major break within the stratigraphical record: a contrast between a primeval world with little biota and a world of abundant and diverse life. The rest of the geological record, about 600 Ma, is best viewed in terms of two halves of a supercontinental cycle. The first half is the story of the assembly of Pangaea during the period from the Cambrian to the Carboniferous. This interval is one of dramatic events, of mountain building and of upheaval and as such is very distinct from the second half of the cycle which commenced with the start of the Permian. The second half is the story of the large Pangaea supercontinent and its dispersal. This time interval has often been referred to as 'the long quiet period' in Europe, since apart from regular transgressions and regressions of the sea and frequent episodes of crustal rifting, it is relatively uneventful. However, it is during this interval that we see the opening of the Atlantic and the separation of the two major continental areas—Europe and North America—in our story.

12

The Precambrian:
Peering into Deep Time

The Precambrian represents an immense interval of time, a deep record of four billion years during which the Earth was created, the oceans and first continents formed, and the first life evolved. In fact, the timescale represented by Precambrian is so vast that it is simply divided into eons (Hadaean, Archaean and Proterozoic) of 1000–2000 Ma duration, each of which represents two or three times the longevity of the more familiar Phanerozoic (Figure 4.8). The great antiquity of the Precambrian has left us with a legacy of rocks which record a complex history of crustal growth, and of multiple episodes of mountain building. Episodes of this story have been written repeatedly in the same rocks, partially destroying and overprinting early records. As a consequence this rock record presents a complex riddle which needs careful research and sophisticated techniques to unravel it. Yet despite this complexity early geologists unlocked some of its secrets using the simple tools of relative chronology (Chapter 3 and Section 7.1.1), and these tools are still applicable to this day.

The origins of the continents in the early days of the Earth have been much debated, and this debate leans heavily on theoretical inferences and deductions made from what field evidence is available. One of the central points of debate is whether or not plate tectonics operated in the Hadaean and Archaean Earth, and if so in what form. What is emerging is the idea that the first continents may have formed at least 4000 Ma ago and that early processes may have been dominated by some form of plume tectonics before about 3500 Ma ago, when the Earth had cooled sufficiently for conventional plate tectonics to become dominant (Figure 10.1). As a consequence

Box 12.1
The Isua Supracrustals: evidence of the ancient world

A fragment of the Earth's earliest stratigraphical record is preserved in the Isua region of West Greenland, 150 km north-east of Godthåb, and has been the subject of detailed stratigraphical investigations since the early 1970s (Moorbath et al 1973). The so-called supracrustals comprise an arcuate belt, some 35 km long and up to 2.5 km thick, of amphibolites, schists, banded ironstones and other rocks which form a 'greenstone' belt surrounding a gneiss dome. The 'supracrustals' are interpreted as metasediments eroded from the early crust, with some volcanics, lavas and intrusions. These metasediments are enclosed and intruded by the polyphase Amîtsoq gneisses, and together they represent a fragment of ancient continental crust, although debate rages to this day about its origins, and the nature of the environments of the metasediments (e.g. Nutman et al 1984; Rosing et al 1996). What is clear is that the age of these rocks is greater than 3700 or even 3800 Ma, derived from Uranium–Lead or Lead–Lead isotopic dating (Moorbath et al 1973; Nutman et al 1984; Nutman and Collerton 1991). This confirms their status as some of the oldest rocks on Earth.

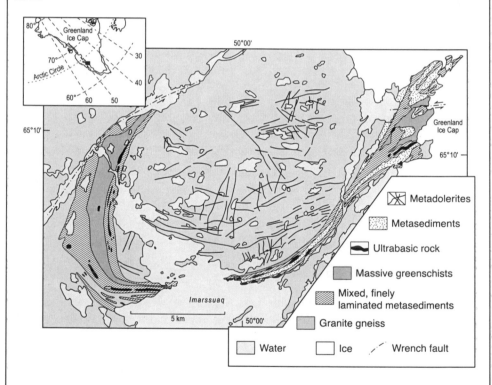

Sources: Moorbath, S., O'Nions, R.K. and Pankhurst, R.J. 1973 Early Archaean age for the Isua Formation, West Greenland. *Nature* **245**, 138–139; Nutman, A.P., Allaart, J.H., Bridgwater, D., Dimroth, E. and Rosing, M.T. 1984. Stratigraphic and geochemical evidence for the depositional environment of the Early Archaean Isua supracrustal belt, southern West Greenland. *Precambrian Research* **3**, 365–396; Nutman, A.P. and Collerson, K.D. 1991. Very early Archaean crustal-accretion complex preserved in the North Atlantic Craton. *Geology*, **19**, 791–794; Rosing, M.T., Rose, N.M., Bridgwater, D. and Thomsen, H.S. 1996. Earliest part of Earth's stratigraphic record: a reappraisal of the >3.7 Ga Isua (Greenland) supracrustal sequence. *Geology* **24**, 43–46. [Diagram modified from Bridgwater et al (1976) In: Escher, A. & Watt, W.S. (eds) *Geology of Greenland*, GGU, Figure 6, p. 23.]

a significant proportion of the Precambrian record can be viewed within a plate tectonic framework.

In the North Atlantic region we have a surprisingly rich legacy of this ancient world, including the oldest known rocks—the Isua supracrustals—from Greenland (Box 12.1). The North American Craton is particularly important and contains an excellent record of Precambrian events since it was welded together in the Early Proterozoic from Archaean microcontinents (Figure 12.1). Since its formation the North American Craton has formed a stable platform against which a range of microcontinents, volcanic island arcs and larger continental bodies have been accreted throughout the Proterozoic. This process of continental accretion culminated in the Grenvillian Orogeny which formed a large continental agglomeration, perhaps a single supercontinent, in the mid-Proterozoic (Box 12.2). At least one earlier agglomeration may have existed, but evidence for this remains equivocal. The Proterozoic supercontinent was to rift apart into smaller continental blocks, before ultimately regathering in the form of Pangaea in the Late Palaeozoic.

In addition to the North American Craton the North Atlantic region contains a smaller, but still significant, cratonic area in Scandinavia (the Fenno–Scandian Shield; Figure 12.1). While both cratons have a coherent story of great antiquity, those areas which effectively came to be sandwiched between them are more difficult to tease apart. This is primarily a function of the numerous orogenic events that have gathered the Precambrian cratons together, and that have deformed the sedimentary basins forming at their margins. This is true of northern Europe and the western seaboard of North America, which has been so heavily deformed and sheared that it simply consists of a collage of terranes. The age, origin, provenance and history of these terranes are notoriously difficult to determine and subject to debate. In light of this we shall first focus on the evolution of the North American Craton culminating in the Grenvillian Orogeny, before zooming-in on two regions which contain complex collages of Precambrian terranes and which feature in later events within our history of the North Atlantic region.

12.1 Evolution of a Precambrian Continent: the North American Craton

The North American craton comprises two parts: that area obscured by the deposition of Phanerozoic sediments; and the exposed, mostly Archaean, rocks of the Canadian Shield (Figure 12.2). The Canadian Shield is one of the most studied areas of Precambrian geology, with a history of participation in it since the nineteenth century. At this time geologists simply used cross-cutting relationships to determine the relative chronology and lithostratigraphy of this vast area, which extends over a large part of the North American continent, including the island of Greenland. For most, the North American craton represents a truly ancient continental area, with the majority of its exposed area being of Archaean age. It is famous for having the oldest recorded rocks on Earth at Isua, in northern Greenland, a snapshot of the Hadaean Earth (Box 12.1). Traditionally, Precambrian facies can be grouped into two broad groups: gneisses of various compositions, but predominantly granitic; and the so-called greenstone belts which contain sedimentary rocks metamorphosed to

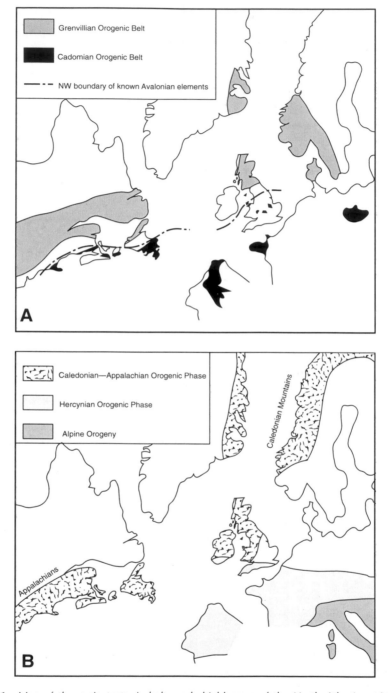

Figure 12.1   *Map of the main tectonic belts and shield areas of the North Atlantic region.* **A:** *Pre-Caledonian elements.* **B:** *Caledonian and post-Caledonian events, note how the Caledonian–Appalachian Orogen overprints the Grenvillian belt, and how the Hercynian Orogen overprints the effects of the Cadomian Orogeny*

greenschist or amphibolite grade, metavolcanics, and interbedded clastic and carbonate metasediments, including banded ironstones. The relationship between these two groups has long been debated, but the most likely explanation is that the gneisses represent early crust, while the 'greenstones' represent volcanic island arcs and associated sediments. As such, recent authors have suggested that the greenstone belts may have formed as volcanic island arcs and are therefore a manifestation of plate tectonic processes. However, the degree of similarity between these ancient volcanic island arcs and those of today is open to some debate.

The dominance of gneisses of Archaean age in the Canadian Shield has led to the idea that most of the primary granitic crust within the shield, which was metamorphosed to give the gneiss, was formed during the Archaean, between 3500 and 2600 Ma ago. However, some older, Hadaean crust is also recorded, particularly in Greenland (Box 12.1). It is now known that this is an oversimplification, and that Proterozoic rocks of similar composition to those of the Archaean exist beneath the cover of Palaeozoic rocks which transgressed the craton during the Cambrian and later on at other intervals in the Palaeozoic (Figure 12.2). In fact the craton is an aggregate of seven Archaean 'provinces' welded by Proterozoic orogenic belts, which provide a record of continental aggradation between 2000 and 1800 Ma ago. The story of this complex area of the Earth's crust is discussed below.

12.1.1 The Archaean Beginnings

It has long been recognised that the Canadian Shield can be divided up into zones or provinces on the basis of the rock suites, and isotopic dating (Figure 12.2). In fact, amongst the earliest uses of absolute dating based on radiometric isotopes (see Section 4.4) was unravelling the huge cratonic area of the heart of North America. Using absolute dating it is possible to identify seven Archaean 'provinces' which are defined as areas of the craton bounded by Proterozoic orogenic belts which have a similarity in the isotopic age range of their constituent facies. Each province itself may consist of several terranes created by collage collision, but are distinct from their neighbours by their overall age range and geographical extent. It is suggested that each of the provinces could represent Archaean microcontinents which were accreted during the Proterozoic. The seven main provinces identified by this method are: Superior, Nairn, Slave, Wyoming, Hearne, Rae, and Burwell.

The Superior Province is the largest of the seven, with an area of 1.6 million square kilometres and comprises a series of east-northeast trending belts of a wide variety of facies, including gneisses and 'greenstones' which probably represent successive island arc terranes accreted together to form larger continents. Four main belt types have been recognised:

1. Volcanic-plutonic terranes, representative of Archaean island arcs.
2. Metasedimentary belts, which have been interpreted by some as accretionary prisms associated with the arcs.
3. Plutonic complexes, perhaps representing small areas of continental crust within the arcs.
4. High-grade gneisses, representing deep eroded remnants of the other three.

Box 12.2
Rodinia or Palaeopangaea? The case for a Late Proterozoic Supercontinent

In recent years, considerable debate has centred on the configuration of a Late Proterozoic supercontinent. The debate has mobilised many different data sets, including polar wandering curves, lithostratigraphy and geochronology, and although most people accept the existence of a Proterozoic supercontinent, debate still continues on the nature of its configuration. Piper (2000) has reviewed two possible configurations. Rodinia — derived from Russian, and effectively meaning the precursor to the continents — takes Laurentia as the 'keystone' and positions other continental areas around it based on their location relative to the Grenvillian orogen (Diagram A). Palaeopangaea is a crescent-like arrangement of continents which is similar to the later Pangaea, with Gondwana as a structural entity, it consists of two other main elements, Atlantica (South America/Africa) and Arctica (Laurentia, Baltica, Siberia; Diagram B). Rodinia has been widely accepted in the literature. However, Piper (2000) has outlined three major difficulties with this interpretation. First, polar wandering data suggests that Laurentia must have separated from the supercontinent by 730 Ma, yet the stratigraphical link is based on post 800 Ma evidence, which apparently limits the duration of any linkage. Second, the transition from Rodinia to Gondwana requires some complex plate movements, with a scattering of the African cratonic elements around the margins of Laurentia. In the face of these data, and on the basis of the simplicity of movement of the continental blocks, Piper (2000) suggests that his Palaeopangaea configuration may be a more viable solution (Diagram B). This retains the continuity of Gondwana during the Palaeozoic, recognises basic stratigraphical similarities and requires a less complex drift

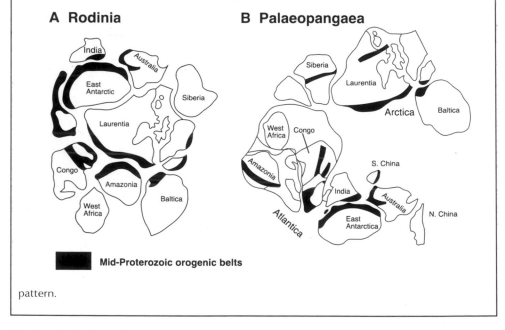

A Rodinia **B Palaeopangaea**

■ **Mid-Proterozoic orogenic belts**

pattern.

Study of the isotopic dates of these belts demonstrates multiple collisions back to 3000 Ma, with progressive assembly of the arc terranes from the north to the south.

The Nairn Province is a triangular block of Archaean crust most famous for containing the oldest known rocks on Earth which date to between 3800 and 3600 Ma ago. As with the Superior Province, the Nairn is considered to have formed by aggradation of individual terranes, each having a different geological history. A

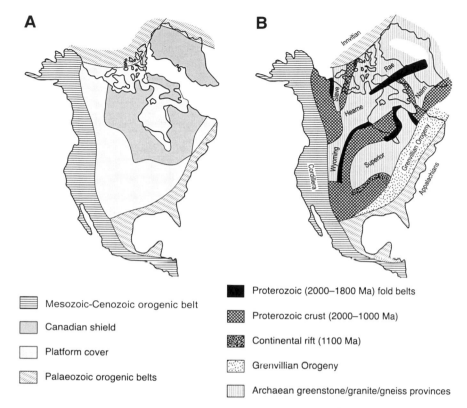

A

B

Mesozoic-Cenozoic orogenic belt

Canadian shield

Platform cover

Palaeozoic orogenic belts

Proterozoic (2000–1800 Ma) fold belts

Proterozoic crust (2000–1000 Ma)

Continental rift (1100 Ma)

Grenvillian Orogeny

Archaean greenstone/granite/gneiss provinces

Figure 12.2   Map of North America showing **A:** The distribution of sedimentary cover, basins and cratons. **B:** The extent of the Precambrian craton. [Based on information in: Hoffman 1989 In Bailey and Palmer (eds) Geology of North America]

range of facies has been recognised, but the province is characterised by its gneisses, and it is possible to identify two generations of gneisses separated by the intrusion of mafic dykes at around 3400 Ma. A suggested origin for these rocks is based on the idea that they represent two generations of supracrustal rocks intruded by granitoid bodies and submitted to repeated metamorphism, separated in time by the intrusion of mafic dykes. The nature of the facies and the complex history of the region mean that no clear scenario can be determined for its formation, although some authorities recognise at least four distinct terranes assembled by faulting between 2750 and 2600 Ma ago.

A small fragment of the Nairn Province can be recognised stranded on the other side of the present-day Atlantic in north-west Scotland (Figure 12.1; see Section 7.1.1 and 12.2). Here granitic gneisses form up to 80 per cent of the Lewisian, and these may be separated into at least two recognisable phases of metamorphism, the Scourian and the Laxfordian, by the intrusion of a mafic dyke swarm, the Scourie dykes, at about 2300 Ma ago. The Scourian rocks have considerable affinity to those of the Nairn Province, and prior to the opening of the Atlantic in the Mesozoic, were in close physical continuity with it (Figure 12.1).

The Slave Province consists of tightly folded, mostly volcanic metasediments ('greenstones') of turbiditic origin with an age range of about 2700 to 2500 Ma. These sediments were intruded by several phases of plutonic intrusions with widely varying lithology, from gabbro to granite. The depositional environment of these metavolcanic turbidites has been widely interpreted over the years. The two most convincing hypotheses involve their deposition either in a basin formed by continental rifting, or adjacent to a volcanic island arc. The contact between the volcanics and older gneisses has historically been interpreted as an unconformity, confirming a rifting hypothesis, but recent examination has determined that it is sheared, and therefore more likely to be a tectonic contact. The absence of facies associated with initial rift valley formation such as evaporites, alluvial-fan deposits and alkali volcanics also tends to count against a rifting origin. A recent interpretation is that the province actually represents the collision of a microcontinent with an island arc, with subsequent obduction of the arc volcanics.

The Archaean rocks of the Wyoming Province constitute only a fraction of the inferred total area of this province, which has been shortened by intense folding and shearing, ultimately complicating its interpretation. For example, it is not known when the province was assembled, and whether it was the result of a collage collision. However, the remnants present do provide evidence for a continental crust older than 3100 Ma in the form of granulite-grade gneisses, together with meta-sediments ('greenstones'), such as quartzites, marbles and pelites, which demonstrate the presence of sand, carbonate and mud facies overlying an inferred continental shelf. Further evidence of shelf sedimentation is provided by the presence of stromatolitic carbonates indicative of shallow marine environments.

The Hearne Province preserves Late Archaean crustal rocks, and comprises low-grade greenschists at its core, with increasing metamorphic grade outwards towards the Proterozoic orogenic belts at the margins. The low-grade core overlies folded volcanic-derived turbiditic rocks of around 2700 Ma in age, penetrated by plutons of varying composition.

The Rae Province is dominated by gneisses with a poorly constrained age of around 2800 Ma. These gneisses surround north-east trending belts of metavolcanic sediments interbedded with quartzites, pelites and iron-formations exhibiting greenschist-lower amphibolite grade metamorphism. Linear calc-alkaline plutonic belts along its margin suggest that the Rae province formed part of an overriding plate to one of the Proterozoic orogenic belts to the north-west. A seventh province, the Burwell Province, comprising mostly granitoid gneisses, occurs towards the tip of Labrador, and has been interpreted as a microcontinental terrane.

12.1.2   Early Proterozoic Orogenesis and Accretion

Each of the Archaean provinces has a distinct age range, a distinct suite of facies, and are mostly bounded by Proterozoic orogenic belts. Each province has a complex history itself, with evidence for a range of Archaean and Proterozoic environments and is often associated with a collage of terranes. It is thought that the Superior,

Nairn and Slave provinces were formed at subducting plate margins as volcanic island arcs, prior to the closure of ocean basins and continental collision. Typically they are associated with accretionary prisms and complex thrust tectonics formed during the creation of these Proterozoic orogens. This contrasts with the other provinces in the hinterland of over-riding plates where continental margin orogens are recorded by complex intrusive volcanics.

The most significant orogen is the Trans-Hudson, which is dated to around 1900–1800 Ma (Figure 12.2). This is the collision zone between the subducting plate containing the Superior Province and the over-riding plate containing the Wyoming and Hearne provinces. Following the accretion of the Archaean terranes which took place between 1950 and 1800 Ma, there was a period of rapid continental growth between 1800 and 1650 Ma, with additional terranes accreted to the Archaean heartland, and the development of the large continental area of Laurentia. This culminated in the Grenvillian Orogeny and the birth of a large supercontinent, which is discussed below.

12.1.3 The Grenvillian: Birth of a Proterozoic Supercontinent?

The Grenvillian is an orogenic episode which took place at around 1100 Ma. Deformation structures associated with this event are identified in a broad belt which extends along the margin of the North American craton and into Scandinavia (Figure 12.1). The Grenvillian Orogeny is still a matter for debate, with some authorities believing that it represented a period of orogenesis associated at one end of a scale with the collision of an island arc with a continental margin, while at the other, the closure of an ocean and the birth of a major supercontinent in the Proterozoic (Box 12.2). The extent of this supercontinent is also a subject for debate, and reconstructing it from the alignment of ancient fold mountain belts such as that provided by the Grenvillian is a difficult business. However, the fact that the Grenvillian Belt can be traced into the Scoresby Sund region of East Greenland, along the south-eastern margin of the North American craton, and into Scandinavia, suggests that at least it involved the coming together of both the Fenno–Scandinia/Baltic shields (Baltica) and the North American Craton (Laurentia). Gondwana may also have been involved in this orogenic event, in which case a large supercontinent similar in composition to that of Pangaea in the Permian may have existed in the Proterozoic. However, determining the location and movement of former continents is based not only on the geometry of former fold belts, but also on palaeomagnetic observations, and polar wandering curves (see Section 5.2). The palaeomagnetic observations for Proterozoic continents are of variable quality and data points are often widely distributed, both in time and space, allowing the data to be interpreted in many different ways and giving a range of different continental configurations and movements.

In the context of the North Atlantic region the close juxtaposition of Laurentia and Baltica is of considerable significance and is the first of two meetings for these cratons. These two cratons rifted and drifted apart in the Late Proterozoic and were separated by the Iapetus Ocean in the Early Palaeozoic, before meeting again during the

Caledonian orogenic phase later in the Palaeozoic, which created the mountains of northern Europe and the Appalachians.

12.2 Life on the Edge: Scotland at the Margin of Laurentia

Between the two main shield areas of Laurentia and Baltica in the North Atlantic region there is a collage of Precambrian terranes caught up in the orogenic belts created as these two cratons collided. Scotland is one such area and contains a number of distinct Precambrian terranes, each with its own stratigraphical record, each providing a distinct 'snap-shot' of parts of the Precambrian (Figure 12.3). The record is often difficult to interpret and the original relationship of one terrane to the next is uncertain however, and as such these areas provide a useful contrast to the detailed and fairly coherent record available for study in the North American Craton.

The Precambrian terranes of Scotland were brought together during the Caledonian orogenic phase, but preserve tantalising glimpses of the Archaean, of Proterozoic environments, and of the Grenvillian Orogeny. In most reconstructions, these Scottish terranes lay off the eastern margin of the North American craton, Laurentia. Scotland has had a complex tectonic history due to its location within the heart of at least two major orogenic episodes, the Grenvillian and the Caledonian, and as such the detailed geological history of the region is difficult to unravel. Despite this, the Scottish terranes provide a fascinating insight into life on the margin of a great Archaean craton (Figure 12.4).

Most authorities accept that three Precambrian terranes can be recognised in Scotland and Northern Ireland, which were accreted into their present position during the later Caledonian orogenic phase in the Palaeozoic (Figure 12.3). These are the Hebridean Terrane, the Northern Highland Terrane and the Grampian (Dalradian) Terrane, each of which is discussed below.

The Hebridean Terrane consists of two distinct stratigraphical elements of very different ages: (1) Lewisian basement and (2) Torridonian. The Lewisian consists of high grade, regionally metamorphosed rocks, mainly gneisses. As discussed above, these rocks are thought to effectively represent a fragment of the North American Archaean granitic crust which may have formed around 2900 Ma ago and which has been subsequently heavily metamorphosed. The Lewisian gneisses have strong parallels with those of the Nairn Province exposed in the Canadian and Greenland cratonic areas. Perhaps surprisingly the Lewisian also contains rocks in which sedimentary structures can still be identified. This shows that the original rocks prior to metamorphism were not simply composed of granitic crustal rocks but also some sedimentary rocks. However, despite these tantalising records we have little evidence of the nature of the developing sedimentary record in the Scottish Lewisian. Some observations can, however, be made from the few sedimentary structures preserved, which show ripple marks indicative of a shallow marine environment. The intensity of metamorphism has been such that no evidence of early life has yet been recovered from these metasedimentary rocks. The development of the Archaean biosphere is therefore unrecorded in Scotland. Evidence from rocks of similar age in Africa and

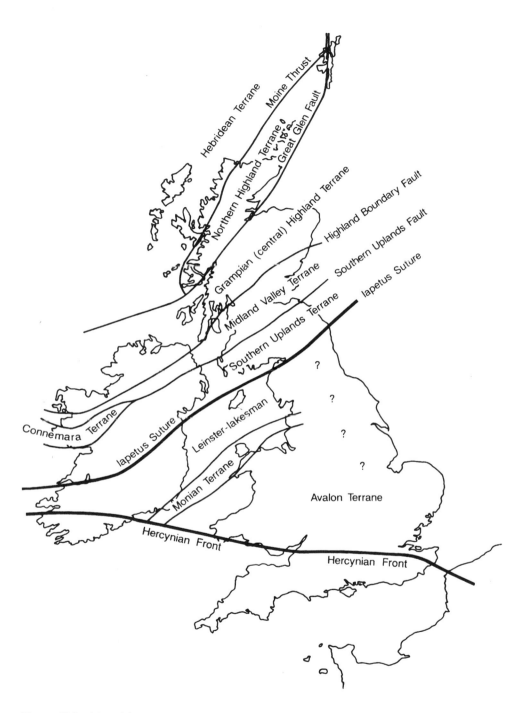

Figure 12.3   *Map of the main tectonic terranes recognised within the British Isles. [Modified from: Bluck et al (1992). In: Cope et al (Ed.) Atlas of Palaeogeography and Lithofacies.* Geological Society Memoir No. 13, Figure 1, p. 3]

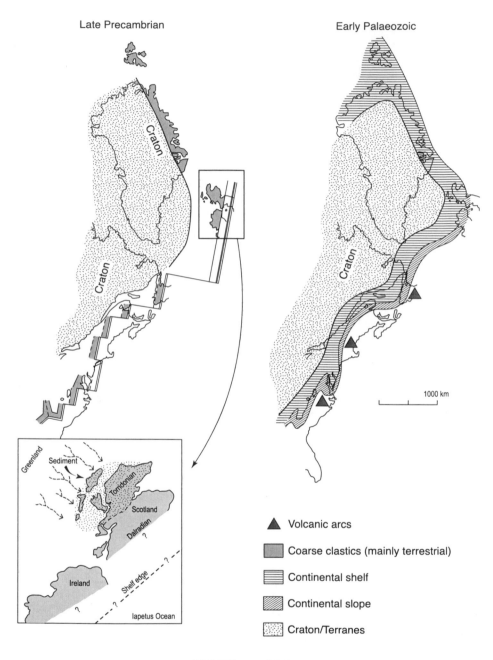

Late Precambrian

Early Palaeozoic

▲ Volcanic arcs

▨ Coarse clastics (mainly terrestrial)

☰ Continental shelf

▨ Continental slope

▨ Craton/Terranes

WESTERN MARGINS

Figure 12.4   *Principal depositional environments along the edge of the North American Craton and along the Avalonian margin. [Modified from: Schwab et al (1988). In: Harris and Fettes (Eds) The Caledonian–Appalachian Orogeny, Geological Society Special Publication No. 38, Figure 1, pp. 76 & 77]*

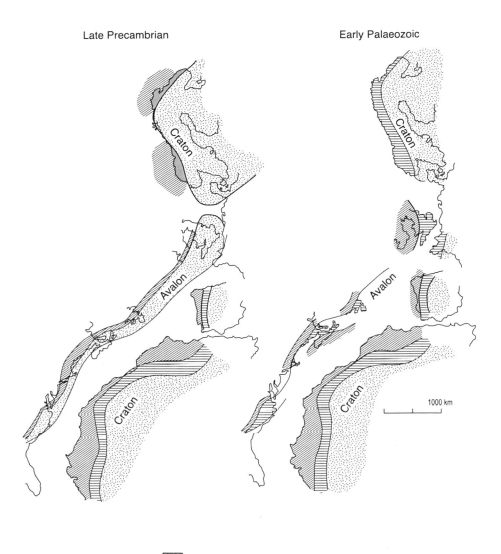

Late Precambrian

Early Palaeozoic

Coarse clastics (mainly terrestrial)

Continental shelf

Continental slope

Craton/Terranes

EASTERN MARGINS

Australia does show, however, that prokaryotic life was developing in an oxygen-poor atmosphere at this time.

The Lewisian records several phases of metamorphism. The first phase of metamorphism probably took place around 2700 Ma ago and produced a diverse assemblage of metamorphic rocks. A second phase of metamorphism has been dated to 2300 Ma, and this remetamorphosed the already heavily altered rocks. The Lewisian has been successfully divided chronologically on the basis of these two phases of metamorphism, which were punctuated by the intrusion of a series of north-west to south-east trending dykes, known as the Scourie dykes. The story of this discovery has already been introduced in Section 7.1.1. The Scourie Dykes are unmetamorphosed by the early phases of metamorphism, but were deformed and altered by the later episode of metamorphism. On this basis two episodes of metamorphism and orogenesis have been recognised: (1) the Scourian (pre-dyke emplacement) and (2) the Laxfordian (post-dyke emplacement).

The Lewisian basement is overlain unconformably by much younger Torridonian sediments which are of Late Proterozoic age. Study of the facies demonstrates that the Torridonian sediments were deposited by huge braided rivers flowing from the Archaean basement of Greenland some 900 to 600 Ma ago, and sedimentary structures within the Torridonian support the idea that the Hebridean Terrane was located on the margins of the North American craton, adjacent to Greenland (Figure 12.4). The basal part of the Torridonian, the Stoer Group, is composed of coarse clastic sediments which buried the eroded landscape of the Lewisian basement and infilled a series of rift basins. These sedimentary rocks are mostly red in colour and this is consistent with an interpretation of their deposition in a hot, arid climate. Subsidence of the rift basins or the evolution of new rifts caused the Stoer Group to be faulted, tilted, eroded and submerged by a shallow sea. Detailed facies analysis shows that the basal units of the younger Torridonian Group were deposited by rivers which prograded into this shallow sea. These units contain simple, marine microfossils, demonstrating the presence of Proterozoic life. Subsequently, deposition became dominated by braided rivers and by large alluvial fans emerging from a highland region to the north-west, located somewhere over Greenland (Figure 12.4).

The Northern Highland Terrane consists of rocks known as the Moines in the north-west Highlands of Scotland. Although these rocks have been heavily deformed and metamorphosed several times, we have sufficient evidence to suggest that they were originally sediments deposited in a shallow marine clastic environment, overlying a Lewisian-type Archaean basement. The age of Moine sedimentation is poorly constrained, but it has been frequently suggested that it may represent an offshore facies of the Torridonian, and therefore be of equivalent age. However, radiometric dating illustrates that this is improbable, since some dates for the metamorphism of the Moine show that this took place during the Grenvillian Orogeny, an event which predates the oldest ages attributed to the Torridonian sediments. The most likely depositional scenario for the original Proterozoic Moine sedimentary facies is offshore, around the margins of the North American Craton, or the continent of Laurentia.

At least two distinct phases of deformation can be identified within the Moine rocks. In the first phase, the original sediments were deformed and metamorphosed during the Grenvillian Orogeny as is evidenced by the tectonic intermixing of Moine metamorphic rocks with the Lewisian basement. This may have occurred due to the collision of Laurentia and Baltica, or following the accretion of an island arc. The second phase of deformation occurred during the Caledonian Orogeny in the Early Palaeozoic, which effectively overprints the effects of the early Precambrian orogeny. In the course of the Caledonian Orogeny the Moine rocks were thrust over the older Hebridean Terrane along a series of thrust planes of which the largest is called the Moine Thrust (Figure 12.3). These thrusts resulted in complex structural relationships which are clearly displayed in the Assynt area of north-west Scotland today.

The rocks of the Grampian (Dalradian) Terrane consist of highly deformed metasediments known collectively as the Dalradian, which were laid down in a variety of depositional environments over a period of 150 Ma, spanning the last part of the Proterozoic. For the most part the Dalradian sediments were deposited in marine conditions within large fault controlled basins associated with a period of crustal stretching along the North American plate margin. The onset of Dalradian sedimentation was probably marked by the transgression of a Late Proterozoic sea across a complex basement consisting of Moine, Torridonian and Lewisian rocks. What is clear is that Dalradian sedimentation postdates the Grenvillian Orogeny, and this was later deformed in the Caledonian orogenic phase.

The Dalradian preserves the record of the Late Proterozoic icehouse as it contains a succession of tillites. These tillites are the product of the widespread glaciation which can be recognised in many parts of the world at the close of the Proterozoic. At Port Askaig, on the island of Islay, glacio-marine deposits are recorded, and these equate with the widespread evidence of glaciation found over many of the Late Proterozoic continents (Box 12.3). The Proterozoic biosphere was dominated by the development of stromatolites, which are consistent with the postulated increase in the level of oxygen in the atmosphere at this time (Figure 10.17). The lower part of the Dalradian sequence contains the remains of numerous stromatolites. The transition to the Phanerozoic marine biosphere dominated by shelly faunas is indicated in the upper part of the sequence, above the tillites, which contains a sparse fauna consistent with that associated with the Early Cambrian faunal expansion (Figure 10.17).

The crustal extension which provided the basins for Dalradian sedimentation changed to compression in the late Proterozoic and this caused the initial deformation and metamorphism of these sediments. This period of deformation, which occurred in several phases between the Late Proterozoic and the Early Ordovician, is usually referred to as the Grampian Orogeny, which is seen by some as the initial phase of the Caledonian Orogeny. The effects of the Grampian Orogeny were felt along much of the eastern margin of the North American plate and may have been caused by the collision of a volcanic island arc, as subduction dominated over spreading in the Iapetus Ocean and this ocean basin began to close.

12.3 Avalonia: a Gondwanan Microcontinent

Avalonia represents a single terrane, a microcontinent that is thought to have initially formed as an island arc obducted against Gondwana, which subsequently rifted from

---

Box 12.3
Snowball Earth

Late Proterozoic successions in many continents indicate that the Late Proterozoic Earth was in the grip of a pervasive glacial episode. Classic tillites are recognised in most of the continents—the first of which was the Port Askaig tillite on the Island of Islay in 1870. It is accepted that the Late Proterozoic glaciation is perhaps the most significant one on Earth, but it is not clear whether it was a synchronous, global event (i.e. an event horizon), or a diachronous response to a series of tectonic episodes associated with the formation and rifting of a supercontinent (Eyles, 1993).

One of the most striking features of this episode of glaciation is the association of the tillites with carbonates. Tillites are frequently interbedded with carbonates and in particular the glacial episode seems to have been terminated by a widespread episode of carbonate deposition, often referred to as 'cap carbonates'. These carbonates have been variously interpreted, as the product of warm-water biogenic precipication through to deposition of detrital carbonate in cold saline water. Recent work has stressed the importance of glacial erosion producing large amounts of carbonate rock flour from uplifted carbonate massifs, explaining many of the interbedded carbonates as detrital deposits. However, some of the cap carbonates are rich in stromatolites and other evidence of a warm-water biogenic origin.

Accompanying the glacial deposits are isotopic excursions in the Carbon-13 record of surface waters of the time, derived from carbon isotopes extracted from the carbonates. The Carbon-13 record shows a decrease greater than any other interval in geological time, and although it can be interpreted in various ways it may imply a reduced level of biological productivity. Palaeomagnetic evidence suggests that ice reached sea level close to the Equator, and a model of global glaciation has been proposed, known as Snowball Earth, in which global glaciation caused a collapse in biological productivity (Hoffman et al 1998). Hoffman et al (1998) suggests that this global glaciation was terminated by an episode of volcanic outgassing, in which atmospheric carbon dioxide rose to 350 times the modern level. In turn this resulted in a rapid transition to greenhouse conditions, and the consequent precipitation of cap carbonates.

The nature of the Late Proterozoic glaciation—global marker event or a succession of episodes—remains uncertain, and resolution of this issue is dependent on better dating control of the glacial strata involved. What is clear is that this period is one of major tectonic, climate and biological change.

Sources: Eyles, 1993. Hambrey, M.J. and Harland, W.B. 1985. The Late Proterozoic glacial era. *Palaeogeography, Palaeoclimatology, Palaeoecology* **51**, 255–272; Hoffman, P.F., Kaufman, A.J., Halverson, G.P. and Schrag, D.P. 1998. A Neoproterozoic Snowball Earth. *Science* **281**, 1342–1346.

---

it during the closure of the Iapetus Ocean. Avalonia is known to have played a major role in the development of the Caledonian orogenic phase, as it converged with Baltica upon Laurentia. The constituents of Avalonia comprise much of Newfoundland, Novia Scotia, New Brunswick and parts of northern Europe, including southern Britain. Discussion of the development of Avalonia is paramount in understanding the development of the North Atlantic region (Figures 12.3 and 12.4).

The Proterozoic basement of Avalonia probably developed as part of the margin of the Proterozoic continent of Gondwana, but its exact relationship is uncertain since it is much younger in age than the West African craton, which has been dated at between 1800 and 1600 Ma. Avalonia consists in most cases of successions of meta-sedimentary rocks, with volcaniclastics of Late Proterozoic age overlain by Early Palaeozoic platform sediments from a continental margin setting. This is typical throughout the Avalon zone, with thick sequences exposed in the Avalon Peninsula of Newfoundland, Cape Breton in Nova Scotia and on the Boston Platform. In England and Wales, small outcrops occur which suggest a development through the gradual accretion of volcanic island arcs, associated accretionary prisms and fore-arc basin sediments on to the edge of the Proterozoic supercontinent of Gondwana, probably in the vicinity of Africa. South of the English Midlands, the Late Proterozoic rocks, exposed in the Channel Islands, northern France, central Europe and northern Spain, are extensively deformed, a product of the Cadomian Orogeny which took place between about 690 and 550 Ma. This orogen preserves slices of ocean crust (ophiolites) trapped by the closure of an ocean basin, and may be correlated with similarly deformed rocks in West Africa. This supports the idea that the Avalonian terrane was sutured to Gondwana through subduction of a Cadomian Ocean in the Late Proterozoic, which may actually have been sandwiched between this and a smaller orogen, centred on Anglesey, that also preserves a suspected Late Proterozoic ophiolite. Whatever its exact provenance, the Avalonian terrane rifted from Gondwana in the Cambrian, and closed upon the North American Craton with the closure of the Iapetus Ocean.

Of particular importance is the occurrence of soft-bodied multiple-celled organisms (metazoans) within the Late Proterozoic shelf rocks of Avalonia (Box 12.4). These organisms were first discovered in the Ediacaran Hills of Australia, and have subsequently been found in Late Proterozoic (Vendian) successions across the world (Section 10.4.2). They are of particular importance because they can be used as guide fossils for the Late Proterozoic, due to their widespread distribution. They comprise a variety of soft-bodied organisms which have been variously ascribed to extinct early metazoan phyla, or related to living organisms, such as cnidarians (corals and their relatives) and worms. *Charnia*, first recognised from Charnwood Forest in midland England is typical of the biota, and has been recognised in other

---

Box 12.4
The Late Proterozoic biota of Avalonia

Since the late 1940s investigators have been successful in finding the delicate impressions of strange quilted organisms which have been collectively named the Ediacara biota (Glaessner 1984). Various modes of life have been suggested for these organisms, and some authorities have related them to Cnidarians, worms and other living phyla (see diagram below). Significantly, Ediacarans have been identified across the globe in rocks of Late Proterozoic age, and are therefore of value in biostratigraphical and biogeographical comparisons, albeit on a relatively crude basis by Phanerozoic standards (Hofmann 1998). Evidence of Late Proterozoic environments in Avalonia is provided by the biotas of Midland Britain and Newfoundland. The Charnian biota of the Woodhouse Beds (Maplewell Group) of Charnwood Forest, England is probably of between 552 and 684 Ma old and contains *Charnia masoni*, *Charniadiscus* and *Cyclomedusa* (Ford 1958, 1980), and similar fossils have been found in Wales (Cope 1977). In Newfoundland, a similar fauna has been described from the Mistaken Point Formation (Conception Group), with an approximate age of between around 590 and 630 Ma (Anderson and Misra 1968, Misra 1969). In both cases, the fossils are preserved in volcaniclastic sediments, and together with the similarity of the faunal elements, it is probable that these now widely separated locations were part of a Late Proterozoic shelf environment on the microcontinental area of Avalonia.

Box 12.4
Continued

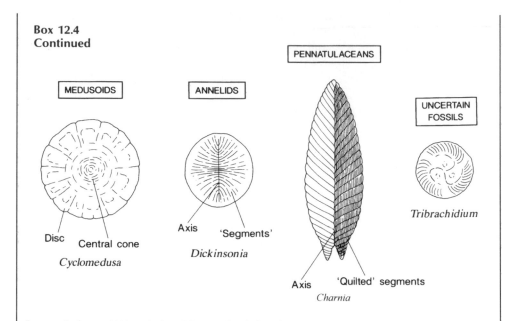

Sources: Anderson, M.M. and Misra, S.B. 1968. Fossils found in the Precambrian Conception Group of south-eastern Newfoundland. *Nature* **220**, 680–681; Cope, J.C.W. 1977. An Ediacara-type fauna from South Wales. *Nature* **268**, 624; Ford, T.D. 1958. Precambrian fossils from Charnwood Forest. *Proceedings of the Yorkshire Geological Society* **31**, 211–217; Ford, T.D. 1980. The Ediacaran fossils of Charnwood Forest. *Proceedings of the Geologists' Association* **91**, 81–83; Glaessner, M.F. 1984. *The Dawn of Animal Life*. Cambridge University Press; Hofmann, H.J. 1998. Synopsis of Precambrian fossil occurrences in North America. In: Lucas, S.B. and St-Onge, M.R. (eds) *Geology of Canada, Volume 7*. Geological Survey of Canada; Misra, S.B. 1969. Late Precambrian (?) fossils from south-eastern Newfoundland. *Geological Society of America Bulletin* **80**, 2133–2144. [Reproduced with permission from Doyle, P. (1996) *Understanding Fossils*, Wiley, Figure 7.4, p. 120.]

parts of Avalonia (Box 12.4). These organisms are the first widespread metazoans, and probably represent improved environmental conditions, and specifically an increased oxygen content of the atmosphere (Figure 10.17). Ultimately, these soft-bodied organisms left no direct descendants to survive into the Phanerozoic, suffering an extinction some time before the close of the Proterozoic. The causes of this extinction are debated, but it has been postulated that increased competition from burrowing organisms, which mark the Cambrian-Proterozoic boundary, may have played a part.

12.4 Assembling the Precambrian Jigsaw

The Precambrian of the present-day North Atlantic region is incredibly rich, including everything from the earliest identified rocks of Earth at Isua in Greenland, through to the officially recognised Cambrian–Precambrian boundary stratotype in Newfoundland, discussed in the next chapter (Box 13.1). The North American Craton is one of the most extensively studied areas of Precambrian crust on Earth;

it has formed a coherent block for at least 1700 Ma, forming the heart of Laurentia, and directly influencing the development of several orogenic cycles. The Fenno-Scandian Shield is smaller, overprinted with Proterozoic orogenic events, but forms the heart of the smaller continent of Baltica. We know of at least one supercontinent in the Proterozoic, and one in the Phanerozoic. In both constructions the ancient cratons form the core of smaller continents which are welded together during orogenesis, and it is in the interpretation of the orogenic belts at their margins that we encounter the complex mosaic of terranes. This is particularly true of the Grenvillian and Caledonian–Appalachian orogens, which are readily identifiable in the North Atlantic region, and centred on northern Europe and the eastern seaboard of North America. Unravelling these terranes has involved generations of geologists, and the discovery of the existence of Avalonia and the deconstruction of the complex of tectonostratigraphical terranes in Scotland, has added to our understanding of how the broad patterns of Earth history have operated. This pattern is further discussed in Chapter 13.

12.5 Summary of Key Points

- **Plate Tectonics.** Very little is known about the global plate tectonic setting during the Precambrian, although it has been suggested by many that there was at least one supercontinental cycle, with the Grenvillian Orogeny signalling the development of a supercontinent. The North American Craton contains abundant evidence of the growth of Archaean crust, and the amalgamation of those into the craton during orogenesis throughout the Proterozoic. Less clear is the amalgamation of the terranes now preserved in the Avalonian region, and it is clear that closure of smaller oceans, such as the Cadomian, may have had a role.
- **Climate.** Our knowledge of the climate of the Precambrian is poor. We do know that the Proterozoic was mostly warm and arid, and this is confirmed in red-beds in Scotland (the Torridonian) as well as in the Avalon Terrane in Newfoundland. The Late Proterozoic, however, was an icehouse interval and abundant evidence of 'snowball earth' is preserved in the North Atlantic region.
- **Sea Level.** Our knowledge of eustatic sea level changes in the Precambrian is poor. Evidence of fluctuation of sea level in the Archaean is suggested by the preservation of sedimentary structures. Definite eustatic episodes in the Proterozoic are illustrated by the onlap of Archaean and early Proterozoic rocks by marine sediments in the later Proterozoic.
- **Biosphere.** The Archaean biosphere was mostly composed of single-celled prokaryotes. The Proterozoic was dominated by stromatolites and much later, by early metazoans. Evidence of all of these have been identified in the Archaean rocks of North America. Importantly, the Ediacaran Biota is identified across the region, demonstrative of its evolutionary success and ultimately, rapid decline in the Late Proterozoic.

12.6 Suggested Reading

Much of the information covered in this chapter is dealt with in more detail by Hoffman (1989) and Rast (1989) for the North American Craton, and Ziegler (1991) for north-west Europe. Detailed discussions of the Precambrian rocks of the North American Craton are examined by Hoffman et al (1990), Reed et al (1990) and Lucas and St-Onge (1998), while Stanley (1989) provides a simple and accessible review. Goodwin (1996) provides a useful overview of the Precambrian of the world. Anderton et al (1979), Craig (1991), Duff and Smith (1992) and Woodcock and Strachan (2000) provide valuable detail from a British perspective, with extensive bibliographies, which is supplemented by the palaeogeographical information of Cope et al (1992). This information can be supplemented for a North European perspective by reference to Ziegler (1990). A useful overview of the Avalonian plate is given by Rast and Skehan (1983). An up-to-date account of the tectonic evolution of the British Isles in the context of northern Europe is provided by Berthelsen (1992).

References

Anderton, R., Bridges, P.H., Leeder, M.R. and Sellwood, B.W. 1979. *A Dynamic Stratigraphy of the British Isles.* Allen & Unwin, London.

Berthelsen, A. 1992. Mobile Europe. In: Blundell, D., Freeman, R. and Mueller, S. (Eds). *A Continent Revealed: the European Geotraverse.* Cambridge University Press, Cambridge, 11–32.

Cope, J.C.W., Ingham, J.K. and Rawson, P.F. 1992. *Atlas of Palaeogeography and Lithofacies.* Geological Society Memoir No. 13.

Craig, G.Y. 1991. *Geology of Scotland.* (Third Edition) Geological Society, London.

Duff, P. McL. D. and Smith, A.J. 1992. *Geology of England and Wales.* Geological Society, London.

Goodwin, A.M. 1996. *Principles of Precambrian Geology.* Academic Press: London.

Hoffman, P.F. 1989. Precambrian geology and tectonic history of North America. In: Bally, A.W. and Palmer, A.R. (Eds). *The Geology of North America – an Overview.* Geological Sociey of America, Boulder, 447–512.

Hoffman, P.F., Card, K.D. and Davidson, A. (Eds) 1990. *Precambrian Craton of Canada and Greenland.* Geological Sociey of America, Boulder.

Lucas, S.B. and St-Onge, M.R. 1998. *Geology of the Precambrian Superior and Grenville Provinces and Precambrian Fossils in North America.* Geology of Canada, Geological Survey of Canada.

Rast, N. 1989. The evolution of the Appalachian chain. In: Bally, A.W. and Palmer, A.R. (Eds). *The Geology of North America – an Overview.* Geological Sociey of America, Boulder, 323–348.

Rast, N. and Skehan, S.J. 1983. The evolution of the Avalonian Plate. *Tectonophysics* **100**, 257–286.

Reed, J.C., Bickford, M.E., Houston, R.S., Link, P.K, Rankin, D.W., Sims, P.K. and Van Schmus, W.R. (Eds) 1990. *Precambrian – Conterminous U.S.* Geological Society of America, Boulder.

Stanley, S.M. 1989. *Earth and Life Through Time.* Freeman, New York.

Woodcock, N.H. and Strachan, R.A. (Eds) 2000. *Geological History of Britain and Ireland.* Blackwell, London.

Ziegler, P.A. 1990. *Geological Atlas of Western and Central Europe.* (Second Edition). Geological Society, London.

13

Pangaea: Birth of a Giant

In this chapter we examine the development of the supercontinent of Pangaea. As we have seen, the formation and ultimate dispersal of supercontinents is a recurrent theme in the crustal evolution of the Earth, and it is widely held that there have been at least three such continents throughout the history of the Earth. The best known is undoubtedly Pangaea, and as it had a dramatic effect on the development of the history of the Phanerozoic, the next two chapters are devoted to it. In this chapter, we examine the events leading to the creation of the continent, and specifically through the development of cumulative orogenies, as ocean basins were created and destroyed in successive Wilson Cycles. At the end of these cycles, the present-day North Atlantic region was central to the new continent of Pangaea. Ultimately the creation of the Atlantic itself was to lead to the demise of Pangaea in the Mesozoic, discussed in Chapter 14.

The construction of Pangaea took place largely through the effects of two cumulative orogenic phases—the Caledonian and Hercynian—which created many of the familiar fold mountain belts of the present North Atlantic region. The Caledonian phase created the Appalachians, the Caledonian mountain chain in Britain, and the mountains of Scandinavia. However, to call it simply the Caledonian would be an over-simplification, and in fact several recognisable orogenic cycles can be identified (Figure 13.1). This is particularly true in the Appalachians, where linear fold belts border that of the earlier Grenvillian Orogeny, and which were deformed through successive orogenies during the Caledonian phase, specifically the Taconic and Acadian. The Hercynian (Variscan) orogenic phase of the Late Palaeozoic created much of the Central European fold mountain belts and again deformed the Appalachians in an event known as the Alleghenian Orogeny, which overprinted much of the earlier events.

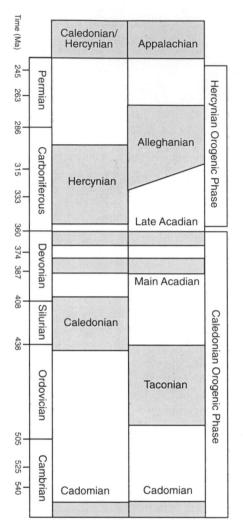

Figure 13.1   *Correlation chart of the main orogenic events of the North Atlantic region during the Palaeozoic*

The Caledonian orogenic phase culminated in the continental collision of three separate components:

1. The early North American continent, Laurentia, sometimes referred to as Laurasia.
2. Baltica, the northern European/Scandinavian continent.
3. Avalonia, the microcontinent comprising northern Europe, Newfoundland and Nova Scotia.

The record of collision is extremely complex, but its most obvious impact was the development of the Caledonian and Appalachian mountain chains in Britain,

Scandinavia and along the eastern seaboard of North America. More significantly, the orogeny led to the creation of a large northern continent that we call Laurussia. Finally, the closure of the Rheic Ocean in the Late Palaeozoic welded Laurussia to Gondwana to form the supercontinent of Pangaea, placing the present-day North Atlantic region at its heart. Ultimately the outcome of these two broad orogenic phases was the docking of the world's continents to form the single supercontinent of Pangaea.

This chapter focuses on the story of the Iapetus ocean, recorded in the North Atlantic region through the rich record of Palaeozoic rocks in Europe, the closure of which resulted in the Caledonian and Appalachian mountain chains. We also review the closure of the Rheic Ocean and the associated formation of the Hercynides over much of Europe.

13.1 'Cornflake' Continents and Cambrian Flooding

We saw in Chapter 12 how the record of Precambrian plate tectonic continents is a complex one and difficult to unravel (Box 12.2). What is clear is that a Late Proterozoic supercontinent had begun to break up and drift apart by the beginning of the Palaeozoic. Palaeomagnetic evidence suggests that this continent rifted and drifted between about 625 and 555 Ma, with the most likely date being around 570 Ma. The dispersal of this supercontinent created a number of small continents based on Archaean cratonic centres, which were distributed across the globe like metaphorical 'cornflakes' in a Palaeozoic breakfast bowl (Figures 10.2 and 13.2).

The rifting and drifting apart of the Proterozoic supercontinent created over 18 000 kilometres of passive margins, providing stable continental shelves which were

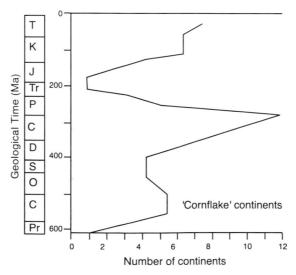

Figure 13.2 *The number of continents during the Phanerozoic [Modified from: Huggett (1997)* Environmental Change, Routledge, Figure 90, p. 90]

extensively flooded in the Cambrian. As discussed in Section 5.1.2, small continents tend to have a lower elevation distribution (hypsometric curve; Figure 5.6) and are consequently more easily flooded than large supercontinents. On the margins of most of our 'cornflake' continents, therefore, we see shallow marine facies overlying both volcanic rocks deposited during rifting, and earlier Precambrian rocks. These facies provide evidence of a marked sea-level rise and consequent trangression in the Cambrian (Figure 10.17). This event had a significant role in the development of Early Palaeozoic environments, and provides an important marker with which to define the beginning of the Phanerozoic, and more specifically the Palaeozoic. The Cambrian transgression was associated with a change from the Late Proterozoic Icehouse to warm greenhouse conditions. The cause of this rise in sea level may have been partly eustatic, associated with deglaciation of Late Proterozoic ice sheets, and partly a result of the subsidence of passive margins created by active rifting as the Proterozoic supercontinent split apart. In either case, this episode of continental flooding created a new ecospace which was to foster the evolution and expansion of the Earth's biota.

The development of new shelf ecospace was a catalyst for the evolution of a new shelly biota, the Cambrian fauna. This fauna was dominated by trilobites, but also included a range of other shelly organisms such as brachiopods, and had forerunners in a diverse group of small shelly fossils (SSF) which replaced the soft-bodied Ediacara biota in the latest Proterozoic. Evidence for this transition from the soft-bodied organisms of the Precambrian to the shelly fauna of the Cambrian is preserved in the stratigraphic records of Avalonia and Laurentia. This transition is also marked by the increase in trace fossil activity, as for the first time, large numbers of organisms burrowed for food and protection in the sea floor (Box 13.1). This was to have an

---

Box 13.1
The Proterozoic–Phanerozoic transition in Newfoundland

The boundary between the Proterozoic and the Phanerozoic, is one of the most important in geology and requires formal designation at an internationally agreed reference section or boundary stratotype section. Until recently, however, there was little agreement as to which locality displayed this boundary in full. Several candidate sections were suggested, but the most viable were two in Russia (in carbonates) and one at Fortune Head on the Burin Peninsula of Newfoundland in siliciclastic sediment. The advantage of the carbonate sections was that they contained abundant Small Shelly Fossils (SSF), a diverse group of organisms which were forerunners of the explosion and diversification of shelly faunas in the Cambrian. However, these organisms are diverse but limited in geographical range. Trace fossils, on the other hand, are abundant at Fortune Head and are more readily found over a wide geographical range, and show a rapid increase in diversity between the extinction of the Proterozoic Ediacaran biota, and the Cambrian explosion. The Burin Peninsula in Newfoundland preserves a thick and largely continuous succession which straddles the Proterozoic–Phanerozoic boundary and as demonstrated below preserves all the main features of the Precambrian–Cambrian transition, with Ediacaran organisms, trace fossils, SSFs and ultimately, the Cambrian shelly faunas (Narbonne et al 1987). The 'golden spike' was symbolically driven into this section in 1994 to mark this important geological boundary (Brasier et al 1994). The boundary was located at the first appearance of a rich assemblage of trace fossils characterised by the feeding trace *Phycodes pedum*. This marker can now be used the world over and geologists can visit its reference locality on this fragment of Avalonia in Newfoundland.

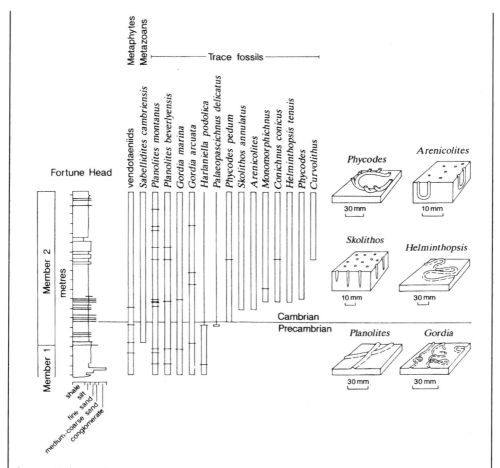

Sources: Narbonne, G.M, Myrow, P.N., Landing, E. and Anderson, M.M. 1987. A candidate stratotype for the Precambrian–Cambrian boundary, Fortune Head, Burin Peninsula, south-eastern Newfoundland. *Canadian Journal of Earth Sciences* **74**, 93–114; Brasier, M., Cowie, J. and Taylor, M. 1994. Decision on the Precambrian–Cambrian boundary stratotype. *Episodes* **17**, 3–8. [Modified from: Narbonne et al 1987 *Canadian Journal of Earth Sciences*, Figure 5, p. 1285.]

important role in the decline of the stromatolites, algal structures which dominated shallow marine environments in the Late Proterozoic, and which were created by delicate algal filaments binding sediments together (Figure 10.16).

The Cambrian faunal development, or expansion, is often referred to as an explosion; an example of a rapid adaptive radiation in which new organisms appeared for the first time, or prospered because of the development of new habitats or ecospace. In addition to the opening of new habitats, the radiation may have been a response to changes in deep ocean circulation, induced by the switch from icehouse to greenhouse states. An abundance of phosphate-rich deposits across the world at this time suggests that there was strong upwelling of deep ocean water, rich in nutrients, during the Early Cambrian. This upwelling may have provided the phosphatic material which was then incorporated into the shells of the first shelly organisms of the Phanerozoic. The development of shells was of great significance and helped fuel the radiation, providing protection for the new organisms from both their physical environments and from potential predators. The rise in sea level and the appearance of the shelly Cambrian Fauna can be recognised on most of the continental margins of the small 'cornflake' continents created by the rift and drift of the Late Proterozoic supercontinent. The isolation of one continental shelf from another led to the development of provinciality within the fossil record; specific fossils come from different continental shelves. This is important in the recognition of ancient oceans, such as Iapetus, and in charting their closure during the successive orogenies of the Palaeozoic.

13.2 The Iapetus Ocean

The Iapetus Ocean was created in the Late Proterozoic and Early Palaeozoic through the rift and drift of the Proterozoic supercontinent. First conceived in the mid-1960s, this ocean was traditionally seen as a 'proto-Atlantic', dividing Laurentia from Baltica and Gondwana. However, this concept has been challenged in recent years, as geologists have attempted to reconstruct the last years of the Proterozoic super-continent. Recent reconstructions consider that the Iapetus Ocean was much more extensive, dividing the large continental masses of Gondwana and Laurentia, rather than simply reflecting an opening of an ocean basin between Laurentia and Avalonia/Baltica, as it has been traditionally portrayed. For example, some researchers have suggested that the ocean formed by the rifting of Laurentia from West Africa, while others have argued that Laurentia rifted from South America. In either case, the end result was the same: that a complex dance of the continents around the Iapetus Ocean led eventually to the collision of Avalonia and Baltica with Laurentia, producing Laurussia. This episode was to leave a lasting legacy in the present North Atlantic region in the form of the Caledonian–Appalachian Orogen.

The existence of the Iapetus Ocean during the Early Palaeozoic was first recognised using fossil evidence. As already discussed, the Early Palaeozoic, and particularly Cambrian faunas are dominated by trilobites which appear to have been confined to the relatively shallow waters of the newly flooded continental shelves. As such, trilobites were unable to cross the Iapetus Ocean, which was deep, requiring trilobites of ocean-going capabilities to cross it. Therefore the trilobite faunas typical of Laurentia are very different from those found in the shelf seas of Baltica and Avalonia (provinciality; Box 13.2). Fossils also provided clues to the evolution of the Iapetus, and it is possible to plot the closure of the Iapetus through time, as these two different faunas became mixed, through the later part of the Early Palaeozoic, as the ocean closed (Figure 13.3).

The closure of the Iapetus Ocean, involving Laurentia, Baltica and the micro-continent of Avalonia, had a dramatic effect on the development of the present-day North Atlantic region via the Appalachian–Caledonian Orogen, the remnants of which are now separated by the present Atlantic. The development of the orogen

Box 13.2
Wilson's trilobites

In 1966 J. Tuzo Wilson published a paper which was to influence the way in which geologists viewed the evolution of the Appalachian–Caledonian orogenic belt. This paper (Wilson 1966) suggested that on the basis of trilobite assemblages, there had been an earlier ocean in the North Atlantic region which had opened and closed during the Early Palaeozoic. Wilson's trilobite assemblages, the Atlantic and the Pacific, were readily distinguished, and yet in the Appalachians and in the Caledonian chain, both assemblages occurred in close juxtaposition. The reasons why the trilobite assemblages were never mixed were obscure in the current continental configuration, particularly as one would have expected that these mobile shelf-dwelling organisms could easily have travelled around in the shallow Palaeozoic seas. The only explanation was that some form of barrier prevented mixing, and the most likely barrier was a deep ocean. Wilson suggested that the line of that ocean was the line separating the Atlantic and Pacific trilobite assemblages. This interpretation inevitably placed northern Britain adjacent to the North American Craton, and a slice of the Appalachians close to southern Britain and Europe, now identified as Avalonia. This influential paper used simple stratigraphical and palaeontological data to reconstruct palaeogeographies, and as such was in the spirit of Alfred Wegener's pioneering work of continental drift in the early part of the twentieth century.

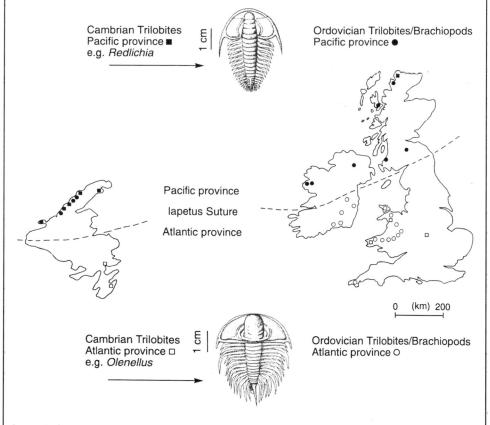

Cambrian Trilobites Pacific province ■ e.g. *Redlichia*

Ordovician Trilobites/Brachiopods Pacific province ●

Pacific province

Iapetus Suture

Atlantic province

Cambrian Trilobites Atlantic province □ e.g. *Olenellus*

Ordovician Trilobites/Brachiopods Atlantic province ○

0 (km) 200

Source: Wilson, J.T. 1966. Did the Atlantic close and then re-open? *Nature* **211**, 676–681. [Modified from: Anderton et al 1979 *A Dynamic Stratigraphy of the British Isles*, Figure 3.5, p. 34]

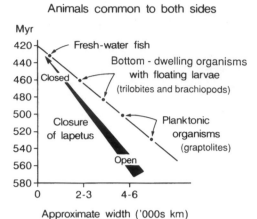

Figure 13.3 *A graph illustrating the progressive closure of the Iapetus Ocean based on faunal evidence. In the early stages of the ocean the only animals that were common to both margins were planktonic organisms (graptolites). With progressive closure of the ocean, a common fauna of bottom-dwelling marine animals with free-floating larvae existed on both margins. Closure of the ocean is indicated by the presence of the same type of fresh-water fish either side of the Iapetus suture. [Modified from: McKerrow & Cocks (1976)* Nature **263**, *Figure 1, p. 305]*

took place over an extended period of 150 Ma, and occurred step-by-step via a range of individual events, such as the accretion of smaller terranes and volcanic island arcs. In order to understand this orogenic phase we must examine events on the opposing margins of the ocean as it closed. Traditional reconstructions refer to these margins as north (Laurentia) and south (Baltica/Avalonia), although it could just have easily have been east and west according to more recent palaeogeographical reconstructions (Figure 13.4). In the face of conflicting opinion we will simply examine the margins of Laurentia and Avalonia, continental areas that were to come together by the end of the orogenic phase.

13.3 The Laurentian Margin of the Iapetus Ocean

Laurentia is one of the most stable of all the Archaean cratons, and has been for at least 1700 Ma. Once the stable cratonic interior was formed by the development of Archaean crust and through Early Proterozoic episodes of crustal accretion, the coherent mass of Laurentia has formed a stable platform against which later continents and terranes were accreted from the mid-Proterozoic onwards. Perhaps the most notable of these was the Grenvillian Orogeny, which is associated with the creation of the Proterozoic supercontinent (Box 12.2). However, the Caledonian orogenic phase followed in the mid-Palaeozoic and created the Appalachian fold mountains, which now lie along the eastern seaboard of North America, adjacent to the Grenvillian Orogen. The Appalachian chain developed first through the

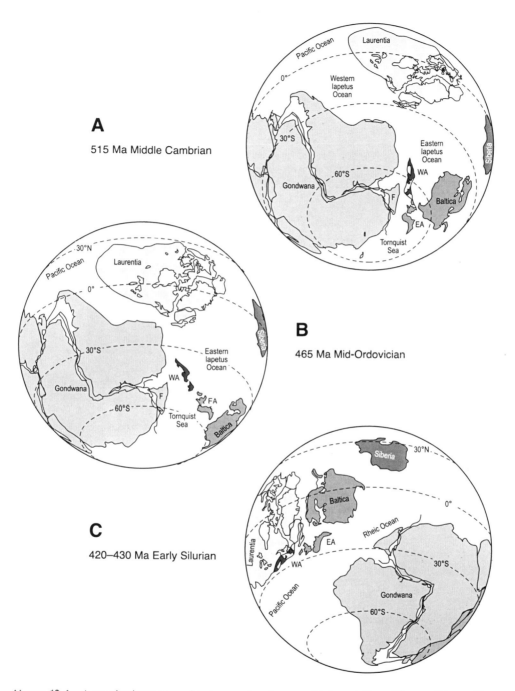

Figure 13.4 *A set of palaeogeographical maps showing recent interpretation of the possible positions of Avalonia, Baltica and Laurentia in the run-up to the Caledonian orogenic phase. WA = Western Avalonia, EA = Eastern Avalonia. [Modified from: Dalziel (1997) Geological Society of America Bulletin* **109***, Figures 14, 16 & 17, p. 37]*

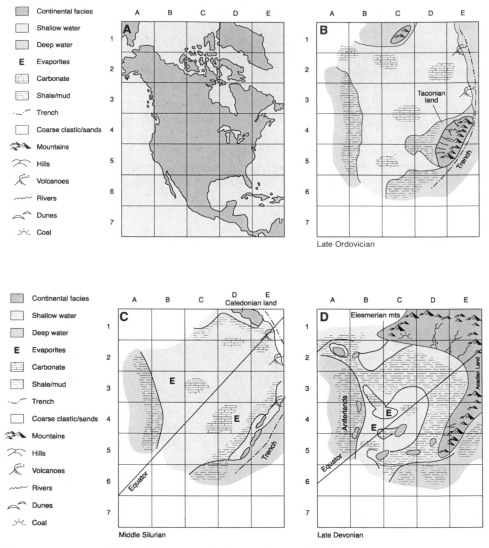

Figure 13.5   *A set of three palaeogeographical maps showing the changing pattern of sedimentation and land in North America during the Early Palaeozoic. [Diagram based on information in: Dott & Batten (1988) The Evolution of the Earth, McGraw Hill, Figures 12.15, 13.16 & 13.32, pp. 318, 351 & 367]*

accretion of a series of volcanic island arc collisions in what is known as the Taconian Orogeny, and was followed by the Acadian Orogeny during which the continents of Baltica and Avalonia docked with Laurentia (Figure 13.5).

However, before the dramatic episodes of mountain building in the mid-Palaeozoic, Laurentia consisted of a stable land mass with an upland interior, with extensive shelf seas around its margins (Figure 12.4). This pattern is reflected today in the thick sequences of Palaeozoic sedimentary rocks which obscure much of the North American Craton (Figure 12.2), leaving only the Archaean upland crust of the

Canadian Shield exposed. The history of the marine inundation of the cratonic margins has become a classic for the study of sequences, or large scale packets of sedimentary rocks bound by unconformity surfaces, representing a pattern of fluctuating sea levels (see Section 5.1.2). Six sequences can be recognised in the North American Craton, each of which has been named after native American tribes (Box 13.3). Importantly, the sequences demonstrate the existence of contemporaneous marine sedimentation on both western and eastern margins of Laurentia, representing continent-wide transgressive events providing therefore a means of intra-continental correlation. The recognition of these sequences was to play an important role in the development of sequence stratigraphy, and are now widely used in framing the geological evolution of North America (Box 13.3).

Two sequences are of direct relevance in considering the pre-Pangaean evolution of Laurentia: the Sauk and the Tippecanoe. The base of the Sauk Sequence is marked by the Cambrian transgressive episode, and the surface over which the sequence was deposited represents a Precambrian landscape of eroded Archaean and Proterozoic crystalline and metasedimentary rocks. The end of the Sauk depositional cycle took place in the early to mid-Ordovician and was associated with a eustatic fall in sea level. This event is coincident with the start of the Taconic Orogeny, part of the Caledonian phase, along the south-eastern seaboard of Laurentia. As a result, along both the Cordilleran and Appalachian margins of North America the basal unconformity surface of the Sauk Sequence is complicated by later orogenesis. Fortunately, however, as sea level rose throughout the Cambrian, it is possible to track the pattern of onlap of the sequence onshore in the continental interior. The exact shoreline is difficult to define but the central upland of the Canadian Shield and the 'Transcontinental Arch'—broadly equivalent to the Proterozoic Trans-Hudson orogenic belt—remained above sea level. The transgressive facies of the Sauk Sequence consists of coarse clastic sediments on the basin's margins, and offshore mudrocks and carbonates, while the regressive facies consist mainly of sands and carbonates. The Burgess Shale, one of the most important fossil lagerstätte, was deposited in deepwater mudrocks during the Middle Cambrian and provides a 'fossil bonanza' which gives a detailed insight into the diversity of Cambrian life (Box 5.6).

The Sauk Sequence is overlain by the Tippecanoe Sequence which is continuous from the Middle Ordovician to the Early Devonian, and as such coincides with the period of closure of the Iapetus Ocean. The Tippecanoe Sequence is deposited upon an eroded surface of the Sauk Sequence, which was formed by a fall in eustatic sea-level in the Early Ordovician. However, in places the distinction between the two sequences is difficult to make, and as is to be expected, a longer history of erosion is identifiable in the cratonic interior than along the continental margins. The facies represented in the Tippecanoe Sequence are again basal coarse clastic sediments, which pass into carbonate rocks. However, deep water, finer clastic rocks replace the carbonates on the eastern margin of Laurentia, due to the presence of subduction and a developing ocean trench along the eastern seaboard, which is associated with the closure of the Iapetus Ocean. The regressive episode which terminated the Tippecanoe Sequence was extensive. The uplift of the eastern margin of Laurentia caused by the docking of Avalonia and Baltica was an important element in this sea level fall and consequent regression, demonstrating the key role of tectonics relative to

Box 13.3
Sequences in the heart of Laurentia

In 1963 L.L. Sloss published a paper which considered the complete stratigraphical record of North America through reference to broad packages of sediment bounded by unconformity surfaces. This approach linked the western and eastern margins of North America chronostratigraphically. He correlated major phases of sedimentation which onlap eroded surfaces in the continental interior to form unconformities, with conformable boundaries at the continental margins. Sloss (1963) was able to draw upon a significant body of work carried out by generations of geologists, but the novelty of his approach was in genetically linking the stratigraphical evolution of North America. The erosion and subsequent burial of each sequence by the next was attributed to continental wide sea level fluctuations which were later correlated with eustatic sea level fluctuations. This paper was to play an influential part in the development of sequence stratigraphy some decades later. Six sequences were identified, covering the whole of the Phanerozoic record, and named after Native American tribes. These sequences provide an essential framework with which to examine the evolution of the North American sedimentary record.

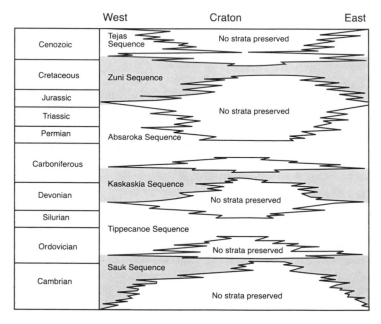

Sources: Sloss, L.L. 1963. Sequences in the Cratonic Interior of North America. *Geological Society of America Bulletin* **74**, 93–114. [Modified from: Sloss, L.L. (1963), *Geological Society of America Bulletin*, Figure 6, p. 110.]

sea level. Finally, the distinction between the Tippecanoe Sequence and the overlying Kaskasia Sequence is easily made due to the widespread nature of the unconformity surface.

It is clear from this discussion that the cratonic sequences can be interpreted on the basis of allocyclic mechanisms, such as eustatic sea level changes (Figure 10.17). The onset of the Sauk and Tippecanoe Sequences can be related directly to eustatic sea level rises, while their termination and the onset of widespread erosion can be related to eustatic falls. The post-Cambrian transgressive event is probably a by-product of global warming. Subsequent changes which define the other boundaries of the Sauk and Tippecanoe Sequences are, however, more likely to be a function of volumetric changes in the ocean basins, caused by variation in the amount and rate of sea-floor spreading, as well as by continental accretion (orogenesis). This is demonstrated in Figure 10.17, as the major transgressive events which create the basal clastic facies of both sequences appear to relate to increases in the rate of ocean crust formation (sea floor spreading), while regressive events are associated with decreases in this rate, as well as with orogenesis.

Smaller scale sea level changes had a dramatic effect on the biosphere at this time, and the Lower Palaeozoic basins of the North American Craton record these. For example, towards the end of the Sauk depositional cycle small scale and episodic extinctions of the dominant trilobite faunas may be related to equally small-scale sea level or climatic fluctuations, although the causes are not entirely clear, and the effects were not widely felt outside of Laurentia. However, a distinct sea-level fall with a much wider impact took place towards the end of the Ordovician, and is closely associated with a glaciation in Gondwana, which was situated in a polar position at that time. This glaciation is unusual in that it existed at a period when the atmosphere contained 16 times as much carbon dioxide, a greenhouse gas, as at present. It has been suggested that this glaciation may have been unstable and subject to rapid variations in both ice volume, and consequently eustatic sea level. The occurrence of icehouse conditions and associated rapid sea-level fluctuations are thought to have been responsible for a mass extinction episode in the Late Ordovician. This led to a severe decline in both benthic and planktonic organisms, and is one of the five main mass extinction episodes of the Phanerozoic (Figure 4.10 and Box 13.4).

13.4 The Avalonian Margin of the Iapetus Ocean

While Laurentia was a large continental area that had remained more or less stable since the mid-Proterozoic, the 'southern' margin of the Iapetus Ocean was in many ways more complex, comprising the ancient continental crustal block of Baltica, and the microcontinent of Avalonia. Although some authorities have suggested that Avalonia is little more than a spur of the Baltica continent, it is actually more likely that it represented a separate entity which came together with Baltica and Laurentia during the Caledonian orogenic phase.

As discussed in Chapter 12, the microcontinent of Avalonia probably commenced life as a volcanic island arc off the coast of the West African Craton. The Avalonian arc was then welded onto the West African Craton by the closure of a small ocean during the Cadomian Orogeny at the termination of the Proterozoic (Figure 13.1), an event which was recorded by deformation in northern Europe. During the Early Palaeozoic, Avalonia, comprising elements of northern Europe, southern Britain and the eastern seaboard of North America, rifted from the southern continent of Gondwana and drifted northwards to form the southern margin of the Iapetus Ocean (Figures 13.4 and 13.6). The original location of Avalonia on the margin of

Box 13.4
The Ordovician Extinction

The Late Ordovician extinction event is in many ways similar to the much more widely studied extinction at the end of the Cretaceous (K–T boundary), in that around 20–25 per cent of marine families were wiped out during the process (Brenchley 1989). Unlike the K–T boundary event, however, it is widely held that the extinction was multi-elemental, and as illustrated below this has been largely associated with the onset of sea level fall associated with the glaciation of Gondwana. Small scale extinctions of trilobites can be identified in the Late Cambrian of Laurentia, and an adaptive radiation of Ordovician organisms is associated with the sea level rise in the Early Ordovician. The initial extinction included nearly 75 per cent of trilobite genera and 25 per cent of brachiopods, and this event can be examined in the carbonates of Laurentia and Baltica. This appears to directly correlate with the onset of a glacio-eustatic fall reducing shelf ecospace for benthic organisms, but global cooling may also have had an effect, as a longstanding greenhouse phase had created stable environments for some time. This has been referred to as the 'first strike' in the extinction event, which reflects rapid environmental change (Brenchley et al 1995). The extinction left an impoverished fauna, the *Hirnantia* fauna, which was more able to withstand cooling. This fauna was later wiped out in the next phase, when sea level rose following the cessation of glaciation, when warm anoxic waters flooded the formerly stable carbonate shelves.

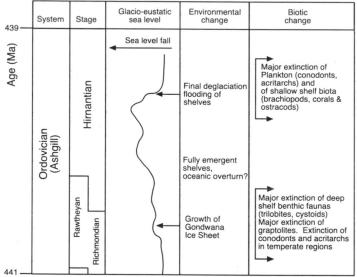

Sources: Brenchley, P.J. 1989. The late Ordovician extinction. In: Donovan, S.K (Ed.). *Mass Extinctions*. Wiley, Chichester; Brenchley, P.J., Carden, G.A.F. and Marshall, J.D. 1995. Environmental changes associated with the 'first strike' of the Late Ordovician mass extinction. *Modern Geology* **20**, 69–82. [Modified from: Eyles (1993) *Earth Science Reviews* **35**, Figure 15.6, p. 124.]

Gondwana, and the precise timing of the rifting is unknown. However, what is clear is that Avalonia was located along the southern margin of the Iapetus Ocean from the Cambrian onwards.

The widespread marine transgression which followed the initiation of greenhouse conditions in the Cambrian along the Avalonian as well as the Laurentian margins of Iapetus, and the sediments deposited are rich in trilobites (Cambrian Fauna). The

Cambrian System was named after rocks in Wales, and these contain abundant evidence of trilobites, particularly *Paradoxides*, one of the most important genera used by Wilson in defining the southern margin of what was later to be called the Iapetus Ocean (Box 13.2). *Paradoxides* can be identified in the Avalonian terranes now incorporated within the Appalachians on the eastern seaboard of North America, and were emplaced there during the Acadian Orogeny (Figure 13.7).

Subduction of the Iapetus Ocean beneath the margin of Avalonia commenced in earnest during the Ordovician (Figures 13.4 and 13.6). This led to the creation of volcanic island arcs in a marine setting, and volcaniclastic arc rocks can be identified throughout the Avalonian terrane (Figure 12.3). In Britain, characteristic volcaniclastics of this age are particularly apparent in North Wales and Cumbria. Subduction ceased for a time in the Late Ordovician and the margin became quiet. Marine deposition occurred in a variety of settings in both shallow and deep water conditions during this period.

The worldwide mass extinction event at the end of the Ordovician (Box 13.4) is identifiable within the Avalonian terranes, and the faunas of the Ordovician carbonates deposited on Baltica at this time. Other important developments in the biosphere also occurred in the Ordovician; in particular the invasion of the land surface by plants. This invasion is suggested by microfossils in Wales, which provide evidence for plant spores. Plant macrofossils (i.e larger fossils), and particularly those of the simple plant *Cooksonia*, are found in Silurian sediments. *Cooksonia*-like plants are actually known from several continents, including South America, and this may add credence to reconstructions showing rifting of Laurentia from the South American continent, sometimes referred to as the Rio de Plata craton, during the Palaeozoic. The arrival of land plants marks a major change in the biosphere as plant life and subsequently the animal life associated with it developed on land in the later Palaeozoic. It would have also had an impact on surface processes, since vegetation today plays an important role in determining the erodability of different surfaces and consequently the rate of sediment supply to depositional basins.

In Avalonia, Baltica and Laurentia, the Silurian is marked by a sea-level rise and a gradual return to a greenhouse world, following the Late Ordovician glacial episode. The greenhouse conditions promoted the growth and development of shelf carbonates, particularly the development of coral and algal reefs which were widespread at this time. This recovery of reef faunas followed the end-Ordovician extinction, as stable shelf environments returned, and promoted reef development on both margins of Iapetus. Alongside the reefs, brachiopod-rich faunas (the Palaeozoic fauna) replaced trilobites as the dominant marine fauna, and are found in a range of marine shelf facies deposited on all margins of the Iapetus Ocean at this time. Ultimately, this pattern of deposition was interrupted towards the end of the Silurian with the closure of the Iapetus Ocean, and the creation of the Caledonian–Appalachian Orogen.

13.5 Converging Margins: the Caledonian–Appalachian Orogen

The evolution of the Caledonian–Appalachian Orogen was very complex. Early interpretations involved the simple linear closure of an ancient ocean (Iapetus

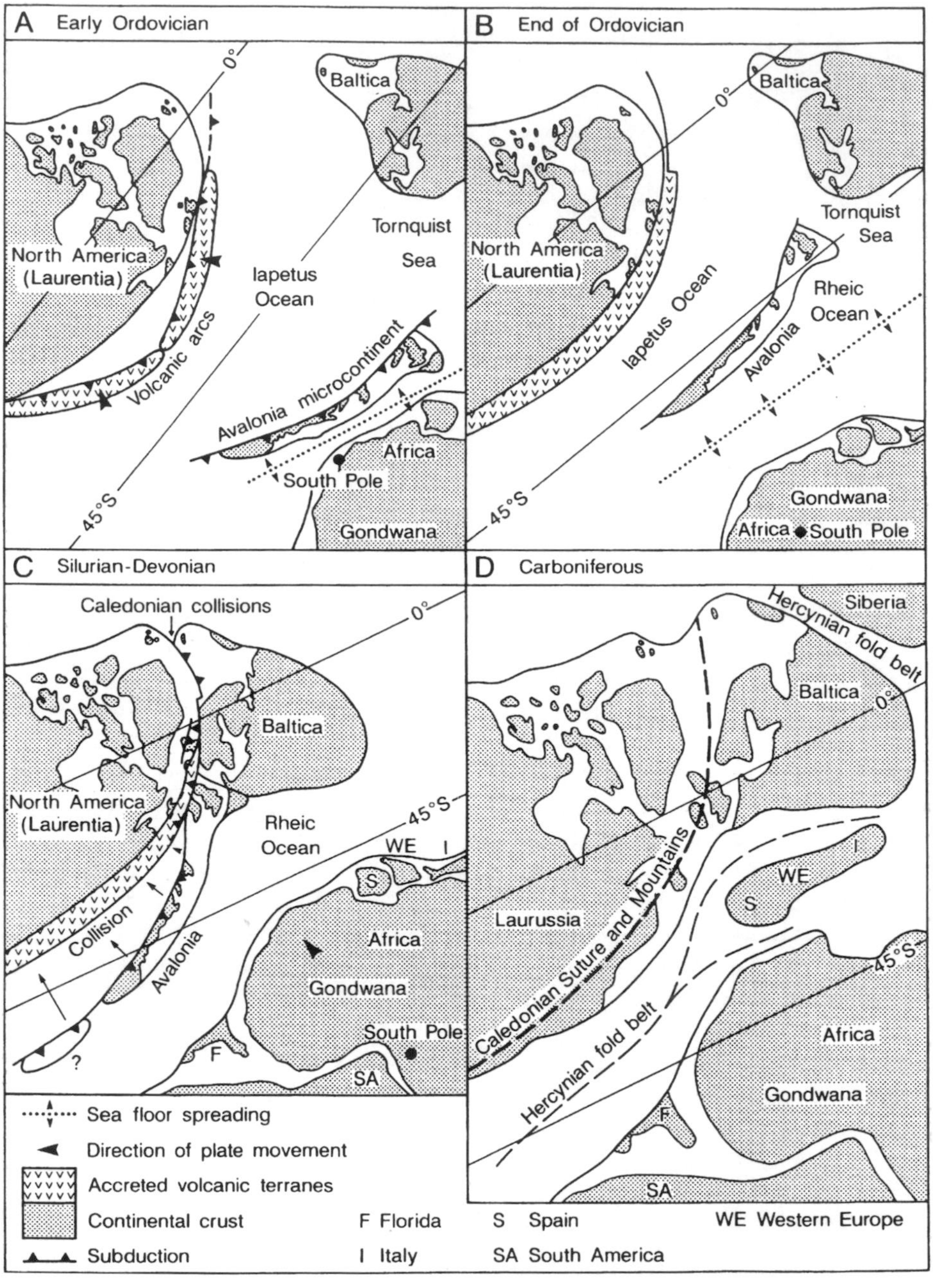

Figure 13.6   *A series of sketch maps of Britain and north-west Europe during the Palaeozoic.* **A, B & C:** *The closure of the Iapetus Ocean and the Caledonian Orogeny.* **D:** *One interpretation of the Hercynian Orogeny.* [Based on information in: McKerrow (1988). In: Harris & Fettes (Eds) The Caledonian–Appalachian Orogeny, Geological Society Special Publication No 38, 405–412; Glennie (1990). In: Glennie (Ed.) Introduction to the Petroleum Geology of the North Sea, Blackwell, 34–77; Scotese & McKerrow (1990) Memoir of the Geological Society of London No 12, 1–21; Berthelsen (1992). In: Blundell et al (Eds) A Continent Revealed: the European Geotraverse. Cambridge University Press, 11–32]

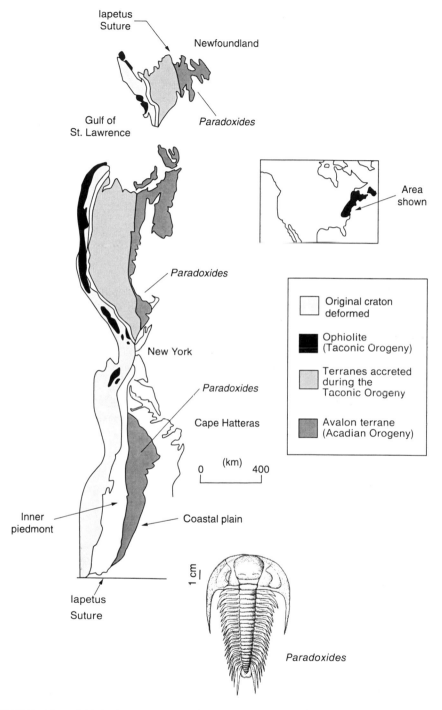

Figure 13.7 *Map of Avalonian terranes on the eastern seaboard of the North American continent.* [Modified from: Stanley, S.M. (1989) Earth and Life Through Time, Freeman, Figure 8.27, p. 223]

Ocean) through subduction. However, it is now clear that collage tectonics played an important role and that a range of different terranes were accreted to the converging margins before the ocean finally closed. The situation is further complicated by the fact that these terranes are allochthonous, a term which refers to the fact that they have been moved from their original place of formation (see Section 7.3.1). Lateral movement of terranes is a common feature of this orogen, as they were 'squeezed' between the colliding plates of Laurentia, Baltica and Avalonia.

On dissecting the Caledonian–Appalachian Orogen it is clear that there are several phases involved associated with the accretion of terranes as the ocean contracted and finally closed. For example, when discussing the Appalachian Orogen, it is usual to consider three main orogenies: the Taconian, Acadian and Alleghenian. Only the first two are associated with the Caledonian orogenic phase, while the last is associated with the later Hercynian orogenic phase that closed the Rheic Ocean, and ultimately led to the creation of Pangaea (Figure 13.1). In northern Europe, it is usual to consider the development of the Caledonian mountain chain as the product of a protracted phase of orogenesis that can be most readily compared with the Acadian events (Figure 13.1). However, orogenic events similar to the Taconian may also be demonstrated from the rock record of Britain. Broadly speaking, then, three phases may be identified in the Caledonian–Appalachian Orogen which are reviewed below.

1. Initial Orogenic Phase. This phase took place some time prior to the Ordovician and involved the collision of volcanic island arcs with the margins of Laurentia. This phase is significant as it includes the development of extensive terranes within the Appalachian belt, and is identified as the Taconic Orogeny in North America. The Taconic mountains form a linear tectonostratigraphical belt adjacent to the Proterozoic Grenvillian Orogen and borders the eastern margin of the North American Craton (Figure 13.5). Collision of volcanic island arcs with Laurentia deformed carbonate platform sediments and shunted the accretionary prism which had been located in the trench over these platform sediments. Related events include the Grampian Orogeny (Figures 13.1 and 13.8), identifiable in Scotland as a probable volcanic island arc collision which was responsible for the first phase of deformation and metamorphism of the Dalradian rocks of the Grampian Highlands. It is important to remember that Scotland was also a component of the Laurentian margin at this time.

2. Terrane Accretion. Following the initial orogenic phase, there was further accretion of volcanic island arcs, back-arc basins, accretionary prisms and fragments of ocean crust known as ophiolites (Figure 13.8). Ophiolites consist of ocean floor basalts, fragments of mantle rock and oceanic sediments. This was probably associated with a reversal in the direction of subduction, with the Iapetus Ocean crust descending northwards beneath Laurentia (Figure 13.8).

3. Continental docking and terrane displacement. On the closure of the Iapetus Ocean the accreted terranes were displaced laterally, due to the oblique collision and docking of Avalonia with Laurentia, which commenced during the Silurian (Figures 13.4 and 13.6). It is this phase which is most often associated with the Caledonian Orogeny in

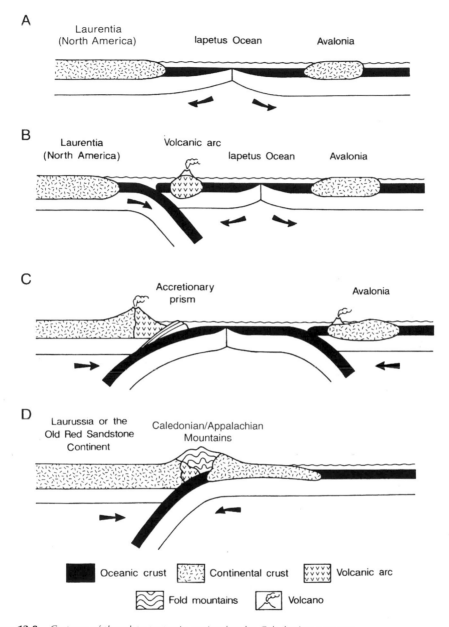

Figure 13.8 *Cartoon of the plate tectonic setting for the Caledonian orogeny*

Europe, and its broadly equivalent Acadian Orogeny in North America. The Avalonian terranes are seaward of the Acadian terranes on the eastern margin of the North American Craton, and are identifiable from the earlier Taconic marine environments by their exotic faunas, most particularly the *Paradoxides*-rich trilobite faunas associated with northern Europe (Figure 13.7).

As a consequence of this complex series of events the history of the Appalachian–Caledonian Orogen is difficult to interpret. Clearly, the juxtaposed terranes we observe today did not necessarily form next to one another and are therefore stratigraphically distinct. This is particularly true of Scotland, which effectively straddles the zone of suturing of Laurentia with Avalonia. Five distinct terranes can be recognised in Scotland (Figure 12.3): the Southern Uplands, Midland Valley, Grampian (Dalradian), Northern Highland, and Hebridean terranes. As befits the classic definition of a tectonostratigraphical terrane, in which differing rock records are separated by distinct faults, these terranes are separated by the Southern Uplands Fault, the Highland Boundary Fault, the Great Glen Fault, and the Moine Thrust, respectively. It is worth while considering these terranes individually as an illustration of terranes generally within the Caledonian orogenic phase.

As discussed in Chapter 12, the Hebridean and Northern Highland terranes are composed of ancient Archaean (Lewisian) and Proterozoic (Moine) rocks which formed part of the leading edge of the Laurentian plate. The Grampian, Midland Valley and Southern Uplands terranes all evolved along different parts of the active plate margin to the east of Laurentia. These terranes were juxtaposed by lateral movement along the axis of the suture during the last phase of the Caledonian orogenic phase.

The Grampian Terrane consists of Dalradian sediments that were deposited in a marine shelf setting on the leading edge of the Laurasian plate. These sediments were first deformed and metamorphosed during the Grampian Orogeny; equivalent in many respects to the Taconic Orogeny identified in the Appalachians. As with the Taconic Orogeny, the Grampian Orogeny may have developed from the collision of a volcanic island arc with the plate margin, although no evidence of a volcanic island arc has yet been found within the Scottish sector of the former plate margin. The Midland Valley Terrane is composed of sediments deposited in fore- and back-arc basins as well as the volcanic products of the volcanic island arc itself. The Southern Uplands Terrane consists of a stacked sequence of sediments of Ordovician and Silurian age that were deposited in a deep-water, marine setting. This sequence has been interpreted as an accretionary prism that accumulated along the subduction zone caused by the descent of the Iapetus Ocean beneath the Laurentian plate (Figure 13.8). The sedimentary facies of continental slope sands and deep water muds is consistent with such a trench environment. However, it has been suggested that a similar sequence of sediments could be produced in what is termed a successor basin. A successor basin is produced along the line of suture between two continental masses. In this case, the successor basin would represent the final remnant of the Iapetus Ocean, although not now floored by ocean crust. Continued compression of the basin by the continents could produce structures reminiscent of an accretionary prism and similar to those found in the Southern Uplands.

Upon the closure of the Iapetus Ocean all three terranes (Grampian, Midland Valley and Southern Uplands) were moved laterally to their present positions (Figure 13.9). At the same time crustal compression and shortening within the leading edge of the Laurasian plate brought the Hebridean and Highland terranes into close

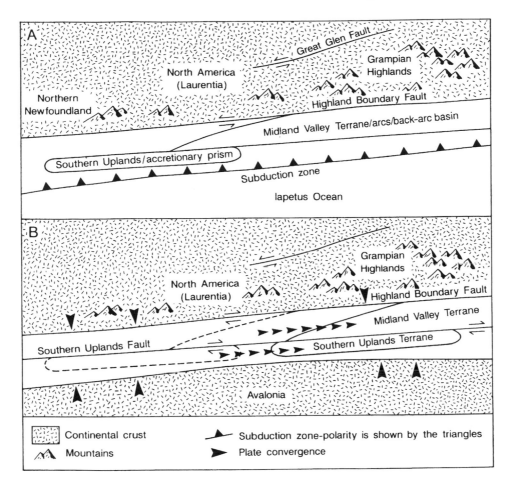

Figure 13.9 *Cartoon to show the lateral movement of tectonic terranes (Midland Valley and Southern Upland terranes) along the Caledonian suture*

contact as the Highland Terrane was thrust over the Hebridean Terrane along the Moine Thrust (Figure 12.3).

13.6 Laurussia: Laurentia, Avalonia and Baltica United

With the closure of the Iapetus Ocean in the Devonian, the continents of Laurentia, Avalonia and Baltica were joined, and a new continent, Laurussia was created (Figure 13.10). The Caledonian orogenic phase culminated at this time with continued compression and lateral displacement along the axis of the suture. Laurussia is often referred to as the Old Red Sandstone Continent. The continent is given its informal name because of the nature of the sediments which were deposited over much of its geographical area in the Devonian. Post-orogenic red-beds were created in the continental interior by the erosion of the Caledonian mountains. The Old Red

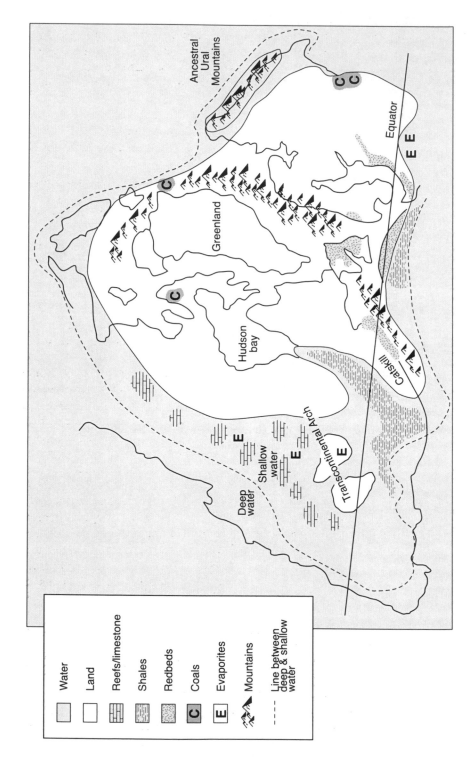

Figure 13.10 *Palaeogeographical map of Laurussia (the Old Red Sandstone continent Late Devonian). [Modified from: Stanley, S.M. (1989) Earth and Life Through Time, Freeman, Figure 13.36, p. 374]*

Sandstone continent experienced a continental semi-arid climate consistent with its relatively high relief following the collision of Baltica and Avalonia with Laurentia. The greenhouse world of the Silurian continued into the Devonian, although climate became more continental as a consequence of the closure of the Iapetus Ocean. Continentality was enhanced by a eustatic sea level fall at the end of the Silurian (Figure 10.17).

Straddling the Iapetus suture zone, Britain experienced the deposition of continental facies (the Old Red Sandstone) from the Late Silurian and by the Devonian, at the culmination of the Caledonian Orogeny, almost all of Britain was land, except for south-west England. In northern Britain, close to the continental interior, coarse sandstones were deposited in basins between mountain blocks (intermontane basins). Drainage in this region was often internal (i.e. not open to a sea), flowing into basins and large lakes developed between mountains. In southern Britain an alluvial plain was crossed by large braided and meandering river systems. The sedimentary record from these areas contains abundant evidence of vigorous water flow and flooding. In some regions, ephemeral lakes and pools formed between the main river course during floods and were quickly desiccated. In south-west England the alluvial plain passed into marine conditions, and this is mirrored with carbonate development along the other margins of the new continent (Figure 13.10). The southern margin of Laurussia south of Britain formed the northern margin of a broad back-arc basin formed by plate convergence to the south of Britain, in northern France.

In the continental heartland of Laurussia, and centred on the old Laurentian interior, the eustatic sea level fall at the end of the Silurian led to widespread erosion of the Tippecanoe Sequence. As a consequence, the basal unconformity below the overlying Karakaskia Sequence is extensive, and befits the scale of the continental collision and uplift that was taking place at the continental margin. Reefs bordered the west coast of the continent, while evaporites formed in shallow basins either side of the Transcontinental Arch (Figures 13.5 and 13.10). Deeper marine sediments are represented by shales east of the Transcontinental Arch, possibly caused by subsidence induced by the loading of the continent by the Appalachian orogenic mountains. These mountains were also associated with a thick wedge of clastic sediments, most probably developed in braided rivers, on the margins of the mountains in the Catskills. As erosion of the Old Red Sandstone continent continued sea level gradually rose, the subsequent transgression depositing the Karakaskia Sequence. Like the Sauk and Tippecanoe sequences beneath it, the Karakaskia Sequence is typically clastic in facies type, although carbonate rocks are more common in the Upper Devonian.

In the sediments of the new continent there is rich evidence of the diversification of the biosphere during the Devonian. Fresh water fish radiated and were widespread in the ephemeral lakes and streams at this time, and the similarity of these fishes during the Devonian compared with invertebrates at other times in the Early Palaeozoic has been used as evidence of the closure of the Iapetus (Figure 13.3). A complex land-based ecosystem of spore-bearing plants (*Rhynia* and *Asteroxylon*) and land-dwelling animals can also be recognised in the fossil record, particularly in the spectacularly preserved fossils at Rhynie, in Scotland (Box 5.6). In East Greenland Late Devonian sediments preserve the remains of *Ichthyostega*, an important link between lobe-finned fishes and later amphibians, which indicates a diversification of vertebrates in the eastern upland region of Laurussia. Elsewhere, the Devonian limestones of the margins of Laurussia preserve evidence of reef faunas, demonstrating the return to carbonate shelves.

A global mass extinction event has been recognised in the Frasnian–Fammenian (end-Devonian), which some researchers have attributed to a short-lived glaciation, despite the presence of widespread greenhouse conditions (Figure 10.17). This event wiped out 60 per cent of the pre-existing benthic fauna, and had a major impact on the reef communities around the margins of Laurussia. However, on land the development of plants and their associated communities appears to have continued unabated. Rapid sea level fluctuations, or changes in ocean temperature and oxygenation appear to be the most likely cause, however, the role of glaciation is likely to have been minimal due to the size of ice bodies which are known to have existed during the Devonian.

13.7 The Final Assembly of Pangaea

The Hercynian orogenic phase led to the final assembly of Pangaea. Again this orogenic phase was associated with several distinct episodes in different locations and is known by a plethora of terms. For example, in Europe it is referred to as either the Hercynian, Variscan or Armorican Orogeny, while in North America it is known as the Alleghenian Orogeny (Figure 13.1). It will be referred to as the Hercynian orogenic phase here, for simplicity. This phase commenced in the Devonian, and its effects were felt throughout the Carboniferous. In effect the orogeny involved the collision of Gondwana (Africa, South America, Antarctica, Australia) with the newly created Laurussia (Laurentia, Baltica and Avalonia), between which a number of smaller microcontinents and associated terranes were sandwiched. Essentially, it involved the closure of the Rheic Ocean which had opened behind the advancing Avalonian microcontinent, and the Gondwanan margin. However, the actual width and significance of this ocean is still open to debate. The line of suture between Laurussia and Gondwana now runs through central France, mirroring the old line of the Cadomian Orogeny of the Late Proterozoic (Figure 12.1).

In Europe, and in front of the advancing continent of Gondwana, a marginal sea or back-arc basin opened during the Early Devonian. This basin, known as the Rheno–Hercynian Basin, opened as a result of crustal rifting behind an island-arc subduction complex in southern Brittany. Floored by ocean crust, this basin was slowly consumed as compression, caused by the collision of Gondwana with Laurussia, forced deformation northwards (Figure 13.11). The style of deformation in the Hercynian orogenic phase was very different from that experienced in the earlier orogenic phase in the Caledonian. The Hercynian phase was mostly completed through thin-skin tectonics. This type of deformation occurs when the upper layers of the crust are uncoupled from rocks deeper in the crust along a décollement surface. This surface acts rather like a polished floor underneath a rug: the rug is easily 'rucked up' in a series of big folds when it is moved quickly over the shiny surface. In central Europe, southern Britain and in the Appalachians the rug is represented by a series of giant overfolds which have been thrust to form giant sheets, called nappes, which advanced northwards during the orogeny. In Britain, for example, the advancing 'front' of deformation finally stopped along a line that runs from South Wales to Kent in south-east England. This line, known as the Hercynian Front, denotes the limit of thin-skin deformation, although the effect of the orogeny was felt in other ways north of the line. In the Appalachians, an Alleghenian Front may also be recognised concentrated in the central and southern part of the belt, and a similar series of nappes over a décollement may be identified. These define the 'Valley and Ridge' terrain of this region, with older blocks such as the Blue Ridge Mountains, being carried 'piggy-back' over the décollement.

Ocean crust from the floor of the back-arc basin was thrust into the pile of deformed sediments and is recorded in south-west England as an ophiolite complex now exposed in the Lizard Peninsula. The final phase of the Hercynian Orogeny was the intrusion of granites of mid-Carboniferous to Permian age. In south-west England a large granite batholith cuts across the complexly folded Hercynian strata. Erosion of these Hercynian uplands has unroofed this batholith to reveal parts of it in the granite upland 'moors' of Dartmoor and Exmoor. Similar granites are known from the southern Appalachians, and represent the termination of the orogenic cycle.

13.8 Carboniferous Environments: Tropical Seas, Luxuriant Forests

During the Carboniferous, and synchronous with the closure of the Rheic Ocean, the continent of Laurussia was extensively flooded. Onlap of the continental interior by marine facies continued throughout the Early Carboniferous, or Mississippian, and carbonates characteristic of the upper part of the Karakaskia Sequence are widespread. The area of Old Red Sandstone deposition close to the Iapetus suture zone was flooded from the south for the first time in the Early Carboniferous (Figure 13.12). This continent-wide transgression can be correlated with the eustatic sea level curve (Figure 10.17), and is associated with a reduction in sea-floor spreading following the Caledonian orogenic phase. Due to the location of Laurussia close to the equator (Figure 11.1) the Early Carboniferous continental seas were warm and tropical, ideal for the deposition of coral reefs and shelly limestones, usually dominated by brachiopod faunas. The Carboniferous limestone facies is persistent over much of Laurussia, and corals and brachiopods are so widespread that they have been employed as guide fossils in Early Carboniferous sequences.

However, towards the middle part of the Carboniferous the global climate reversed into an icehouse state (Figure 10.17). This created a global fall in sea level as water was locked into a series of ice sheets in Gondwana, which were to wax and wane for over 100 Ma (Figure 10.5). However, away from Gondwana, direct evidence of this glaciation is limited. Despite global refrigeration, much of Laurussia, and particularly the part we are considering, was still experiencing tropical humid conditions due to its equatorial location at this time (Figure 11.1). Direct evidence of this location can be found in the presence of bauxite deposits in Scotland dating from this time interval, indicating warm, humid equatorial conditions.

A major eustatic sea level fall did take place at the end of the Early Carboniferous (Mississippian). Translated into a regressive phase, this effectively terminated the Karakaskia Sequence on the North American Craton. With regression, deltaic and fluvial river systems prograded over the carbonates in Laurussia. The return to clastic sedimentation was probably caused by renewed uplift of the Caledonian and Appalachian mountains, either as a result of the release of intra-plate stress generated

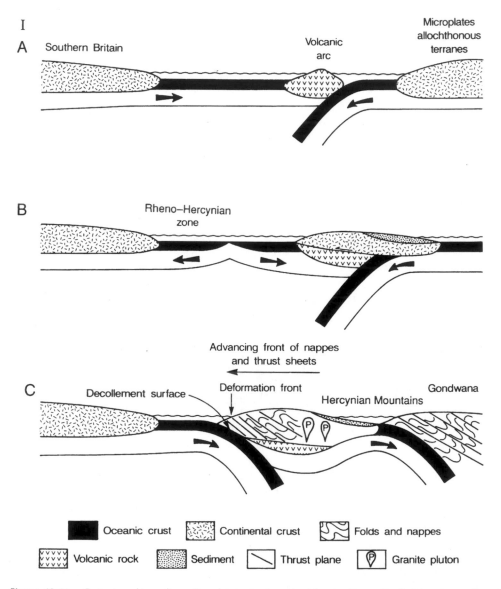

Figure 13.11   Cartoons of two alternative plate tectonic settings for the Hercynian Orogeny in southern Britain. [Based on information from: Berthelsen (1992). In: Blundell et al (Eds) A Continent Revealed: the European Geotraverse. Cambridge University Press, 11–32; Dunning (1992). In: Duff & Smith Geology of England and Wales, The Geological Society of London, 523–561]

during the Caledonian orogenic phase, or as a result of the compressional regime of the developing Hercynian orogenic phase to the south (Figure 13.12).

In fact, by the end of the Early Carboniferous, there was regular interchange between carbonate and clastic sediments, and this is developed as a regular alternation of these facies. Patterns of cyclic sedimentation such as this are commonly referred to as cyclothems. In northern England the cyclothems are known as the Yoredale Cycles, and in this case may actually have been caused by local tectonic

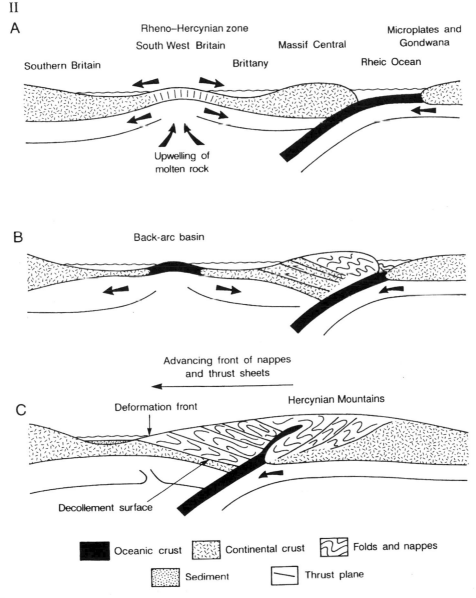

Figure 13.11   *Continued*

adjustment of horst blocks within a block and basin topography within the continental interior caused by the transmission of stress from the continental margin (far-field stress). Here, uplift of horst blocks caused local regressions and the progradation of deltas, while depressions ultimately led to local transgressions, and with a consequent return to carbonate deposition.

In the continental interior of North America, the Karakaskia Sequence was replaced by the Absaroka Sequence, with the marked regressive phase at the termination of the Karakaskia Sequence (Box 13.3). The Absaroka Sequence actually records many small scale fluctuations in relative sea level, and is characterised by clastic deposition in many small scale basins, replacing the predominantly carbonate deposition of the previous sequence. In Britain, the effects of the sea level variation are also felt in northern England which was covered by coarse clastic sediments known as Millstone Grit in deposits up to two kilometres thick. These gritstones are periodically interrupted by marine bands; shales formed by short-lived transgressions (marine incursions), and these provide event horizons with which to correlate the sequence. These marine incursions have been related to eustatic sea level fluctuations because of their widespread occurrence and some authorities have implicated variation in the Gondwanan ice sheets as a possible explanation. However, the size of this ice sheet is unlikely to have been much bigger than the present-day Antarctic Ice Sheet. If this were to melt it would only cause the sea level to rise by 65 metres, yet some of the recorded Carboniferous sea level fluctuations are larger than this. The number of fluctuations appears to have decreased through time, and the intensity of fluvial deposition also declined, such that in the Late Carboniferous (Pennsylvanian) there was a marked increase in tropical vegetation on the tops of the deltas and along the river flood plains.

If the Early Carboniferous (Mississippian) of the Laurussian continent was marked by carbonate deposition consistent with a tropical location, then the later Carboniferous (Pennsylvanian) is associated predominantly with the luxuriant coal forests which created the coal deposits of northern Europe and Pennsylvania. The forest vegetation of the Late Carboniferous consisted for the most part of lycopod spore-bearing plants, which underwent a major radiation in the Late Devonian and Carboniferous (Figure 10.17). Although restricted to swampy environments because of their reproductive cycle, these plants were able to build large trees for the first time, with the development of vascular woody tissue. On their death, the remains of these lycopods and other spore-bearing plants accumulated as peat in the humid swamps of this large deltaic region. Compression of this peat has produced the coal on which the industrial wealth of western nations was originally built. Rare, uncompressed coal balls formed by the accumulation of carbonate in pockets, thereby preventing compression, produce a Lagerstätte which provides an important glimpse into the detail of the lycopod structure. The coal is found in association with lacustrine muds and coarse clastic sediments introduced by local fluctuation in the sea level. Periodically minor marine transgressive episodes record the presence of continued sea level fluctuations, but local tectonic movements appear to be the principal control on sedimentation within the coal swamps.

The Carboniferous sequences of much of Laurussia were to be subsequently deformed during the Hercynian orogenic phase, and gave way to the post-orogenic

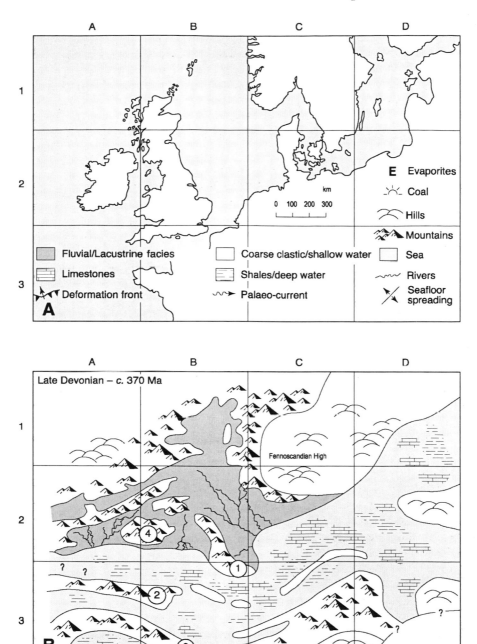

Figure 13.12   *A set of palaeogeographical maps showing the changing depositional environments within Europe, located at the heart of Laurussia during the Late Devonian and Carboniferous. [Based on information from: Ziegler (1990) Geological Atlas of Western and Central Europe. Geological Society Publishing House; Cope et al (Ed.) (1992) Atlas of Palaeogeography and Lithofacies. Geological Society Memoir No. 13]*

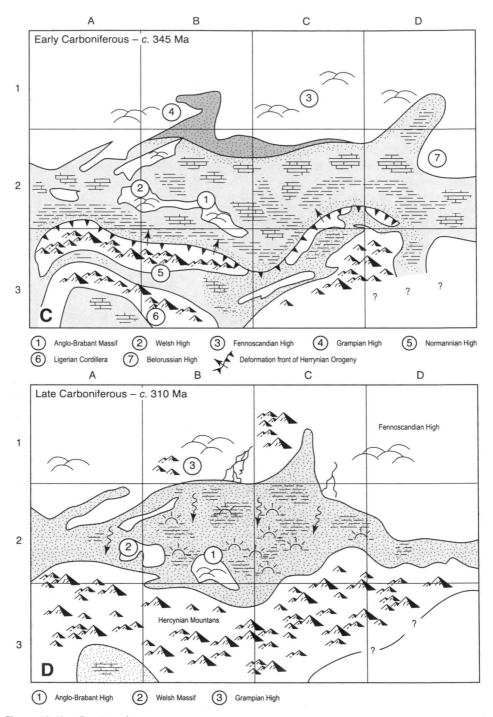

Figure 13.12   Continued

sedimentation of the Permian. At the heart of a supercontinent, the present day North Atlantic region experienced widespread desert conditions, and as a consequence, the rich floras and faunas of the Carboniferous were replaced by the mostly arid facies of the continental interior, tempered with carbonate deposition in small basins and marginal seas.

13.9 The Birth of a Phanerozoic Giant

Several supercontinental cycles are thought to have existed, most in the eons of time encapsulated in the Precambrian and the geological record of the Phanerozoic is dominated by a cycle associated with the development of Pangaea. Commencing with dispersed 'cornflake' continents, cumulative orogenic episodes, a succession of Wilson Cycles gathered them together again to create the Phanerozoic giant. The present-day North Atlantic region is central to our understanding of these events, as although now separated by the Atlantic itself, the Caledonian–Appalachian Orogen provides clues as to how the closure of the 'proto-Atlantic' Iapetus, took place. Closure of this ocean basin united Avalonia and Baltica with the essentially Precambrian craton of Laurentia, and placed northern Europe close to North America as we know them today. From British, Irish and Newfoundland perspectives this orogenic phase was also to unite their component geographical parts for the first time, and from the Devonian onwards a coherent geological history was built in these areas. Eustatic sea level rise enabled the widespread deposition of carbonates in epicontinental seas.

Closure of the Rheic Ocean to the south of Laurussia resulted in widespread deformation across the southern margin of Laurussia, and ultimately created Pangaea. Uplift associated with the developing orogen changed the emphasis away from carbonate deposition through to deltaic sediments, and it is in these locations that luxuriant coal forests developed. Ultimately orogenesis was to create the arid continental interior of Pangaea, replacing the rich Carboniferous environments with deserts. The aftermath of this orogenic phase and the continued development of Pangaea is discussed in the next chapter.

13.10 Summary of Key Events

- **Plate Tectonics.** During the early part of the Palaeozoic the Earth's continents began to move together in the construction of the supercontinent of Pangaea. This resulted in a major orogenic phase, known as the Caledonian, as the Iapetus Ocean closed as part of this overall continental aggradation. The process, which started with the Caledonian orogenic phase, was completed by the collision of Gondwana with Laurussia to form a large supercontinent (Pangaea). This collision resulted from the Hercynian orogenic phase.
- **Climate.** Throughout this period, global climate was in a greenhouse state, with the exception of a brief period of icehouse conditions during the Ordovician. Although polar ice caps grew in Gondwana, located over the South Pole, the effects of this

global cooling were not pronounced in Laurussia, which in part was experiencing deep water sedimentation. The return to greenhouse conditions following the Ordovician Icehouse event is recorded by the deposition of terrestial red beds during the Devonian. Global climate was in a greenhouse state which was reversed towards icehouse conditions during the Carboniferous, when polar ice sheets again developed over Gondwana. This interval of global cooling is not well-recorded in the North Atlantic region, due to its equatorial position at this time.

- **Sea Level**. From the Cambrian onwards eustatic sea level rose rapidly and peaked in the Early Ordovician when the growth of the polar ice caps resulted in a sea level fall. Sea level continued to rise during the Silurian, with the return to greenhouse conditions. The general sea level rise during this period resulted in the deposition of shallow marine sediments over much of Avalonia and Laurentia. After the Silurian, sea level fell. At the start of the Carboniferous, eustatic sea level rose, a trend that was reversed with the growth of polar ice sheets in the latter part of the Carboniferous.

- **Biosphere**. The start of the Cambrian saw a rapid radiation in shelly marine organisms, known as the Cambrian explosion or expansion. At the end of the Ordovician a major extinction event decimated the Cambrian Fauna, which was replaced by a brachiopod-rich fauna, the Palaeozoic Fauna. The Cambrian and succeeding Palaeozoic Faunas are represented in the sedimentary sequences of the northern and southern ocean margins. The invasion of the land took place in the Ordovician; this is recorded in the sedimentary sequences of the region. A Late Devonian extinction event wiped out many benthic and reef species, associated with global cooling. The early part of the Carboniferous saw the development of widespread carbonates consistent with icehouse conditions and high eustatic sea level. The Carboniferous interval saw the radiation of tree-forming lycopods. These formed in the swamp conditions created by the sea level fall in the later Carboniferous. Both elements are well-represented in the North Atlantic region.

13.11 Suggested Reading

Much of the information covered in this chapter is dealt with in more detail by Anderton et al (1979), Craig (1990), Duff and Smith (1992) and Woodcock and Strachan (2000) for Britain, and Bally (1989) and Rast (1989) for North America. Information on the plate tectonic palaeogeography of this period is provided in Bond et al (1984) and Dalziel (1987). Ziegler (1990), Berthelsen (1992) and Cope et al (1992) contain a series of detailed palaeogeographical maps for the British Isles, which are accompanied by excellent summaries and provide a vital source of information. Whittaker (1985) provides information on the sedimentary basins of England and Wales located at the heart of the Laurussian continent during the latter part of the period reviewed in this chapter. North American perspectives are given in the good general texts by Dott and Batten (1989) and Stanley (1989), as well as in the more advanced collections of papers edited by Sloss (1988) and Hatcher et al (1989). Finally, Embry et al (1994) represent a comprehensive collection of papers on all aspects of the geology of Pangaea.

References

Anderton, R., Bridges, P.H., Leeder, M. R. and Sellwood, B.W. 1979. *A Dynamic Stratigraphy of the British Isles*. Allen & Unwin, London.

Bally, A.W. 1989. Phanerozoic basins of North America. In: Bally, A.W. and Palmer, A.R. (Eds), *The Geology of North America: an Overview*. Geological Society of America, Boulder, 397–446.

Berthelsen, A. 1992. Mobile Europe. In: Blundell, D., Freeman, R. and Mueller, S. (Eds). *A Continent Revealed: the European Geotraverse*. Cambridge University Press, Cambridge, 11–32.

Bond, G.C., Nickeson, P.A. and Kominz, M.A. 1984. Breakup of a supercontinent between 625Ma and 555 Ma: new evidence and implications for continental histories. *Earth and Planetary Science Letters* **70**, 325–345.

Cope, J.C.W., Ingham, J.K. and Rawson, P.F. 1992. *Atlas of Palaeogeography and Lithofacies*. Geological Society Memoir No. 13.

Craig, G.Y. 1991. *Geology of Scotland*. (Third Edition) Geological Society, London.

Dalziel, I.W.D. 1997. Neoproterozoic–Paleozoic geography and tectonics: review, hypothesis, environmental speculation. *Geological Society of America Bulletin* **109**, 16–42.

Dott, R.H. and Batten, R.L. 1988. *Evolution of the Earth*. (Fourth Edition). McGraw-Hill, New York.

Duff, P. McL. D. and Smith, A.J. 1992. *Geology of England and Wales*. Geological Society, London.

Embry, A.F., Beauchamp, B. and Glass, D.J. (Eds). 1994. *Pangea: Global Environments and Resources*. Canadian Society of Petroleum Geologists Memoir 17.

Hatcher, R.D., Viele, G.W. and Thomas, W.A. (Eds) 1989. *Appalachian–Ouachita Orogen in the United States*. Geology of North America, Volume F2, Geological Society of America.

Rast, N. 1989. The evolution of the Appalachian chain. In: Bally, A.W. and Palmer, A.R. (Eds). *The Geology of North America: an Overview*. Geology of North America, Volume A, Geological Society of America, Boulder, 323–396.

Sloss, L.L. (Ed.) 1988. *Sedimentary cover–North American Craton: US*. Geology of North America, Volume D2, Geological Society of America.

Stanley, S.M. 1989. *Earth and Life Through Time* (Second Edition). Freeman, New York.

Whittaker, A. (Ed.) 1985. *Atlas of Onshore Sedimentary Basins in England and Wales: Post-Carboniferous Tectonics and Stratigraphy*. Blackie, Glasgow.

Woodcock, N.H. and Strachan, R.A. (Eds) 2000. *Geological History of Britain and Ireland*. Blackwell, London.

Ziegler, P.A. 1990. *Geological Atlas of Western and Central Europe*. (Second Edition). Geological Society, London.

14

Pangaea and the Opening of the Atlantic

The supercontinent of Pangaea was to exist for some 200 Ma, following its final assembly during the Hercynian orogenic phase. This giant continent exerted a strong influence on the nature of the sedimentary facies that were laid down around its margins and in its interior during this period. Our area of study, the North Atlantic region, had a central position within the supercontinent, and the interplay of sedimentary environments in this region illustrates the interplay of climate, sea level and the biosphere. The first part of the interval following the assembly of Pangaea, Permian to Recent, is influenced by events resulting from the after effects of the continent's tumultuous construction. The second part of this interval is, however, dominated by the opening of the Atlantic and the break-up of Pangaea, as well as the impact caused by the final closure of the Tethyan Ocean in southern Europe (Figure 10.2 F–I).

14.1 Geography of a Supercontinent

The sheer size of the supercontinent of Pangaea is perhaps difficult to grasp, as is the concept that all the world's continents were gathered together in a single mass, surrounded by an ocean that made up the majority of the Earth's surface (Figure 10.2 F–I). This ocean, often referred to as Panthalassa, actually has its remnants in the Pacific today, as the opening of the Atlantic at the centre of our region required an consequent increase in subduction on all sides of what is known today as the Pacific

Rim. As commonly reconstructed Pangaea had its long axis oriented north–south, such that there was land from the equator to almost the poles, and this is reflected in the orientation of the majority of the continents today which rifted from Pangaea during and after the Mesozoic.

To the east of the continent lay the Tethys, an ocean which started as a broad gulf or embayment but which was to undergo successive phases of rifting and spreading. The Tethys effectively divided Pangaea into two major crustal components based on ancient continental groupings: the northern continents, or Laurasia (effectively Laurussia plus Asia), and the southern continents, or Gondwana. These continents are effectively hinged in the west, and opening with the changing fortunes of the Tethys in the east. Complete marine links from the Tethys westwards between these two blocks were created at sea level high stands in the Jurassic, and later during rifting. At least three successive incarnations of the Tethys are interpreted from the geological history of Asia, with small microcontinents rifting away from the southern margin of the Tethys, and drifting northwards before colliding with continental crust north of it. Successive waves of microcontinents made this journey, which has been described metaphorically as the motion of 'windscreen wipers' across the Tethys, and these slivers were important in building up the Tibetan Plateau, and other parts of south-east Asia as a complex collage of allochthonous terranes. The most well-known of these slivers was the Cimmerian microcontinent, which collided with Asia in the Jurassic. Tethys was to be finally closed in the Palaeogene with the docking of the Indian sub-continent with the rest of Asia (Figure 10.2 G–H).

The sheer size of Pangaea meant it straddled the equator and reached to the poles. Plant fossils contribute to our understanding of the geological evolution of Pangaea, with recognisable floral zones equating with climatic belts. During the Mesozoic Greenhouse, Pangaea supported seasonally adjusted forests close to the poles. From the Late Palaeozoic onwards, this huge continent also supported the free movement of large land animals, and some were effectively cosmopolitan, such as the mammal-like reptile *Lystrosaurus* in the Late Palaeozoic, and the dinosaurs in the Mesozoic. The development of marine faunal distributions was also strongly influenced by this giant, with many organisms capable of movement from the Tethys around the southern cape of Pangaea and into the Western Cordillera. A more restricted fauna existed in the more restricted regions of the north, and this distinction into a huge Tethyan marine realm and a smaller Boreal one is undoubtedly a function of the geography of Pangaea.

14.2   The Atlantic at Birth; the Beginning of the End for Panthalassa

Construction of Pangaea was largely complete by the end of the Carboniferous, and in many areas of the continental interior the compressional stress derived from the Hercynian orogenic phase led to the gradual inversion of the Carboniferous basins, such that areas which were formerly depositional basins now became gentle rolling uplands subject to denudation. A new tectonic style was also developing. Extensional tectonics, on a regional scale, dominated the Permian throughout the North Atlantic region. These tensional stresses within the crust resulted in a series of rift basins, particularly in south-eastern North America, and in the North Sea. There are two possible explanations for these tensional stresses. One possibility is that they were caused by the release of intraplate stress generated during the Hercynian orogenic phase. The alternative explanation is that they may have been caused by a forerunner of the tensional regime which was to open the Atlantic and begin the break-up of the supercontinent in the Cretaceous.

In south-eastern North America, the Newark Rifts, a series of cratonic rift basins which parallel the main North Atlantic rift zone, developed in the Triassic (Figure 14.1). This phase of rifting was accompanied in the eastern United States by basaltic intrusions, of which the Palisades, a thick basaltic sill, is a good example and is well exposed in the Hudson River in New Jersey. These continental rifts were mostly filled with red-beds of Triassic age, and subsequently by lake deposits similar to those in the East African Rift Valley today. Like the East African Rift system today, the Newark Rift valleys supported a rich fauna, and these Mesozoic lake sediments are famous today for their diverse trace fossil assemblages of dinosaur footprints.

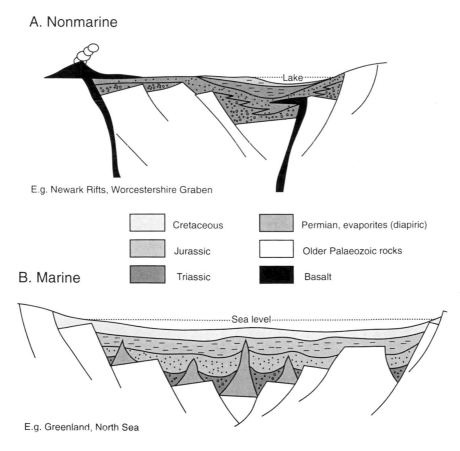

Figure 14.1   *Stylised models of sedimentation within Late Palaeozoic and Mesozoic rifts. [Modified from: Dott & Batten (1988)* The Evolution of the Earth, *McGraw Hill, Figure 16.4, p. 465]*

In the North Sea, two major rift basins, the Southern North Sea Basin and the Northern North Sea Basin were created in the Permian, separated by a raised block known as the Ringkøbing–Fyn High which was later cut by two north–south trending grabens (Central and Viking Grabens) (Figure 14.2). These rift valleys were infilled initially by non-marine Permian and Triassic sediments before being flooded as the rifts widened with the opening of the Atlantic. This phase of flooding is recorded by marine sediments of Jurassic and Cretaceous age. Other rift valleys can be identified within the North Atlantic region. For example, the Labrador Sea, between Greenland and North America, opened at this time as did the Mississippi Embayment, along which the current Mississippi river flows (Figure 14.3). Perhaps most notable, however, was the opening of the Gulf of Mexico, which started to separate the North and South American continents bound together in Pangaea. As with the North Sea, the initial phase of rifting brought with it the deposition of non-marine sediments, followed by marine sedimentation as the rifts widened. In the case of the Gulf of Mexico, close to the Equator, marine sedimentation was actually in the form of extensive evaporites, which were replaced by more open marine conditions by the end of the Jurassic. This region was to be an important migration pathway between the Pacific and Tethyan oceans.

It is possible to suggest, therefore, that this period of tensional rifting represents the birth pangs of the Atlantic Ocean. Irrespective of this it had a major influence in the development of the sedimentary facies of both the European and North American epeiric seas (continental seas) which are a feature of the Jurassic and Cretaceous successions within our North Atlantic region.

Also influential during this time is the development of the orogenesis on the western margin of the North American continent. From the Carboniferous onwards, the western margin of North America was subject to a series of collage collisions associated with subduction of the Pacific plate beneath the North American sector of Pangaea. In broad terms, this process reduced the ancient Panthalassa Ocean in size, and this continues today, with volcanism and seismic activity being common along the western margin of the North American continent, part of the 'ring of fire' that demonstrates the subduction of the Pacific on both of its margins. This is perhaps most famously represented in the Cascades of western North America, where Mount St Helens erupted so spectacularly in the early 1980s.

Subduction and consequent collage accretion of arcs and microcontinents onto the western margin of the North American craton has taken place almost continuously since the Carboniferous. Despite this, it is possible to recognise a number of distinct orogenic phases when accretion took place on this margin. The first of these is the Antler Orogeny, in the latest Devonian and early Carboniferous (Figures 13.5 and 14.4) which led to the collision of the Klamath Arc with the western margin of the continent in Nevada. This collision is the first of many that created the collage of terranes which make up the Western Cordillera of North America (See Section 7.3.1). As with terranes associated with later collisions, the Klamath Terrane had its own geological history and fauna prior to docking, and a subsequent history of lateral movement along its bounding faults during later orogenic phases. The Antler Orogeny was followed in Nevada by the Sonoma Orogeny in the Late Permian, which saw the compression of a small oceanic basin against the continental margin, and the welding of the Sonoma Terrane to the Pacific margin (Figure 14.4).

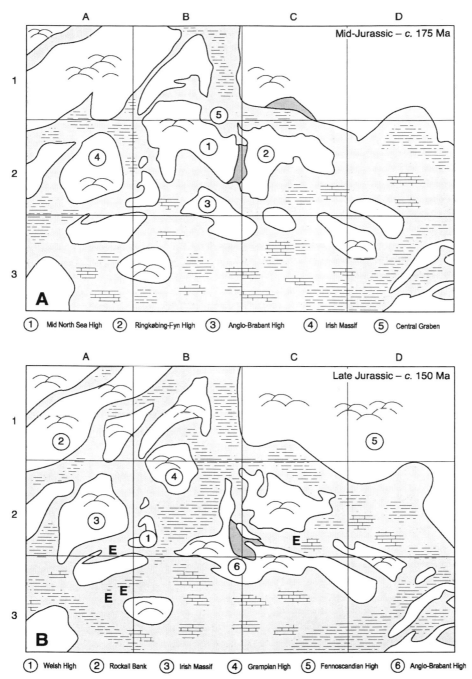

Figure 14.2 *A set of four palaeogeographical maps showing the changing depositional environments within Europe, during the Late Jurassic and Cretaceous see Figure 13.12 for locational grid and key. [Based on information from: Ziegler (1990) Geological Atlas of Western and Central Europe. Geological Society Publishing House; Cope et al (Ed.) (1992) Atlas of Palaeogeography and Lithofacies. Geological Society Memoir No. 13]*

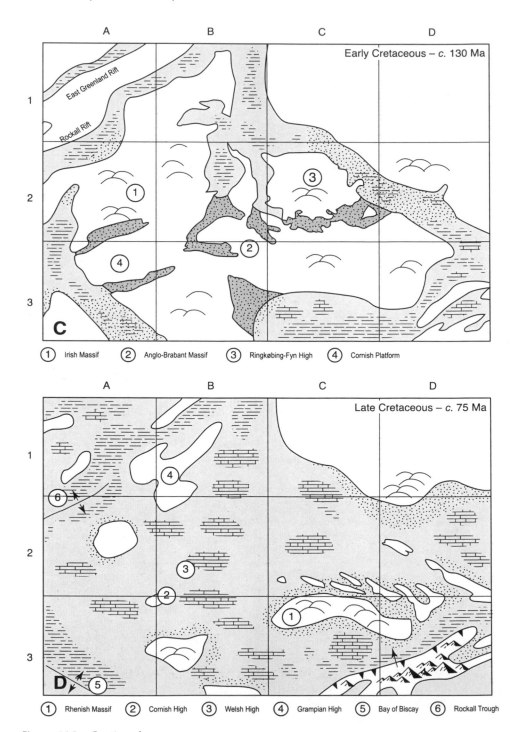

Figure 14.2 *Continued*

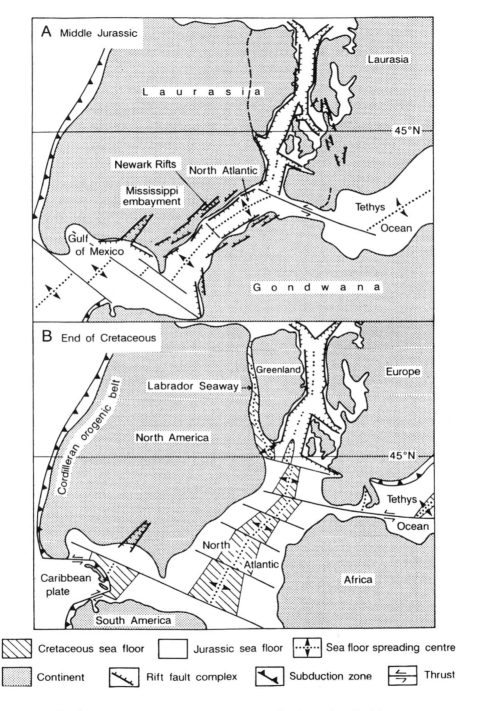

Figure 14.3 *Sketch maps showing the opening of the North Atlantic. [Modified from: Dott & Batten (1988) The Evolution of the Earth, McGraw Hill, Figures 16.2 & 16.3, pp. 463 & 464]*

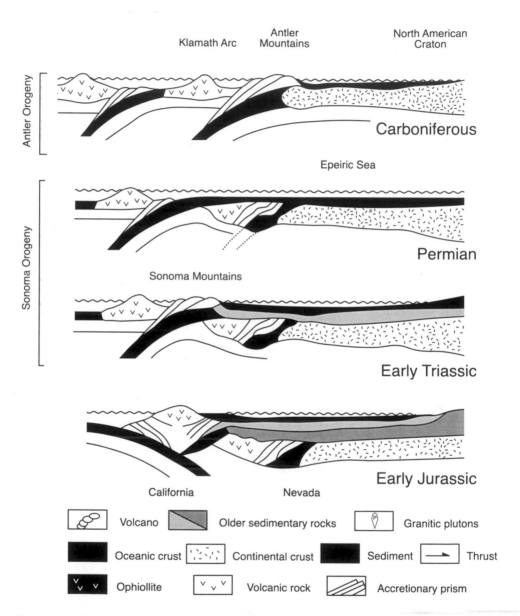

Figure 14.4   *Plate tectonic cartoons for the evolution of the Western Cordillera in North America. [Based on information from: Dott & Batten (1988)* The Evolution of the Earth, *McGraw Hill, Figures 16.7 & 16.29, pp. 469 & 489; Stanley, S.M. (1989)* Earth and Life Through Time, *Freeman, Figure 15.43, p. 465]*

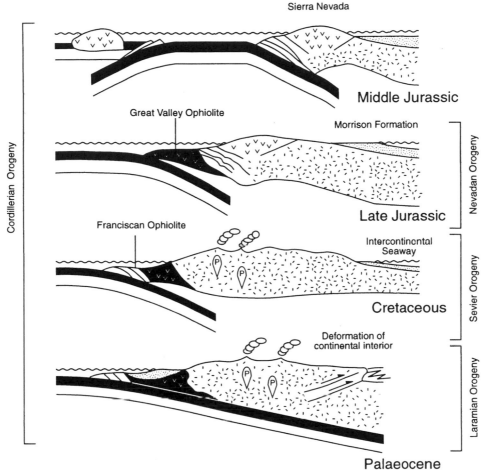

Figure 14.4   *Continued*

The Mesozoic history of the western margin of the North American craton follows a similar pattern to that set at the end of the Palaeozoic, with successive collage collisions as subduction of the Pacific plate continued. Termed the Cordilleran Orogeny, the successive accretion of a complex of terranes, including the Great Valley and Franciscan ophiolites, began in earnest and spanned the Late Jurassic to Early Cenozoic interval; working out its geological history is complex (Figure 14.4). As we have seen in Section 7.3.1, unravelling the complex history of terrane accretion recorded in the Western Cordillera has played an important role in developing our understanding of the process of orogenesis. Some authorities recognise the successive accretionary events as separate orogenic episodes in their own right, with major episodes occuring in the Late Jurassic (Nevadan), Late Cretaceous (Sevier) and Early Cenozoic (Laramian) (Figure 14.4). In all cases, the episodic record of accretion had a dramatic impact on the deposition and preservation of marine sediments in the cratonic interior, as discussed below.

14.3 'The Long Quiet Period': Permian to Cretaceous Depositional Environments

With the closure of the Rheic Ocean during the Hercynian orogenic phase, Pangaea became a physical entity, and was to have dramatic influence on Earth environments for some 200 Ma. Following the cessation of major tectonic activity, the pattern of sedimentation became more passive, and was dominated by the interplay of climate, sea level and biosphere. However, as we have seen, this passive depositional period was complicated through the impact of uplift and orogenesis due to the onset of Atlantic rifting and the development of the Cordilleran mountains in North America. This section examines all of these factors and considers the pattern of post-Hercynian sedimentation and the development of the epeiric seas characteristic of the Mesozoic Greenhouse world.

A direct result of the Hercynian orogenic phase was the production of a broad belt of mountains which constrained patterns of sedimentation and provided a source of clastic sediments. In North America, the Hercynian or Alleghenian Orogeny was responsible for renewed tectonic activity in the Appalachians, which overprinted the earlier Caledonian (Taconic and Acadian) orogenic phase (Figure 13.1). It was also responsible for the creation of the Ouachita fold mountains in south-eastern North America. In Europe, the Hercynides (Variscides) are an important feature of the central portion of the continent, and the Hercynian deformation front may be identified as far north as southern England and Wales (Figure 12.1). In all cases, however, the wider influence of the orogenic phase was the uplift and rejuvenation of older fold mountains, which provided a new landsurface for erosion. As a result, the fold mountains of the Hercynian orogenic phase and the rejuvenated mountain belts helped constrain marine basins, and provided a major source of clastic sediments.

With the closure of the Rheic Ocean the North Atlantic region records a slice of geological history at the centre of the supercontinent of Pangaea. This story is dominated by sedimentation in a continental interior, which was flooded periodically by epi-continental or epeiric seas. Epeiric seas are marine environments founded on continental crust. Their area is determined by both land elevation and by fluctuations in relative sea levels caused by changes in climate, tectonics and in the volume of the world's ocean basins. The transgression and regression of these seas was a major feature of the Mesozoic pattern of sedimentation in the North Atlantic region. In this section we examine first the post-orogenic sedimentation patterns of the Permian and Triassic as the North Atlantic region became part of Pangaea, and second the history of epeiric seas on the margins of Pangaea up to the birth of the Atlantic Ocean.

14.3.1   Deserts and Marginal Seas: Permian and Triassic Pangaea

The Permian depositional environments of the North Atlantic region were largely determined by its position in the centre of the Pangaea supercontinent, but also by the prevailing global climatic state (Figure 10.17). As discussed in Section 14.1, Pangaea encompassed all regions from high to low latitudes, and the pattern of floral development within the supercontinent reflects this. In high latitudes floras (including the famous seed-fern *Glossopteris*) adapted to cooler environments dominated Gondwana in the south and Siberia in the north. In low latitudes tropical floras, responsible for the coal swamps of the Late Carboniferous and Early Permian, were dominant. Our present-day North Atlantic Region was just north of the equator, and therefore the flora and pattern of sedimentation were dominated by warm continental conditions. This is in stark contrast to the global record of icehouse conditions at this time (Figure 10.17).

For some 30 to 40 Ma much of the North Atlantic region was part of a desert at the heart of Pangaea which was comparable in size to that of the Sahara today (Figure 14.5). The Early Permian saw extensive desert weathering of the Carboniferous basement and the deposition of sediment in large dunefields. Large playas or desert lakes existed between the intensely weathered uplands created by the Hercynian orogenic phase, and were frequently desiccated to give salt pans. Marine conditions existed on the fringes of the continent, especially so on the western margin of the North American continent, although marine facies were mostly absent from the cratonic interior (Figure 14.5). This is reflected in the distribution of the Absaroka Sequence, the deposition of which commenced in the Late Carboniferous over an extensive erosional surface created by the Hercynian uplift (Box 13.3). The Absaroka Sequence dominated the marine basins of North America until the Jurassic, when the onset of the Cordilleran orogenic phase caused a widespread regression.

A small but notable exception to the deposition of non-marine sediments in the cratonic interior of North America, is the Late Permian Capitan Reef complex which developed in a small basin (Delaware Basin) west of the Ouachita mountain belt (Box 14.1). The Delaware Basin was a small epeiric sea almost at the heart of the continent which remained connected to the seas of its western margin. It retains its basin form today, and is valuable in its contribution to the study of carbonate systems and Palaeozoic reefs, but represents a special case in an otherwise continental interior location. Marine conditions were also experienced in northern Europe during the Late Permian, with the development of the Zechstein Sea, a warm shallow saline sea in which carbonate-rich rocks were deposited. This extended from the North Sea region eastwards, and bordered the Hercynian mountain chain in central Europe (Figure 14.2). As with the Delaware Basin, the shoreline of the Zechstein Sea was subject to periodic movement and the deposition of carbonate rocks within it was periodically interrupted by the deposition of evaporites and clastic sediments. In fact, the sediments along the margins of the Zechstein Sea have a cyclic nature, the cycles recording the alternation between carbonate, evaporite and clastic sediments. These local sea-level fluctuations probably reflect local tectonic movements and are not believed to be eustatic in nature.

The close of the Permian is associated with the most dramatic mass extinction in Earth history, in which 50 per cent of marine families became extinct (Figure 4.10). This event marks one of the major chronostratigraphical boundaries, that of the Palaeozoic–Mesozoic boundary. Most studies suggest that this extinction may have been caused by a dramatic fall in sea level which reduced the area of shelf sea and increased the climatic continentality of Pangaea, but an alternative is that the

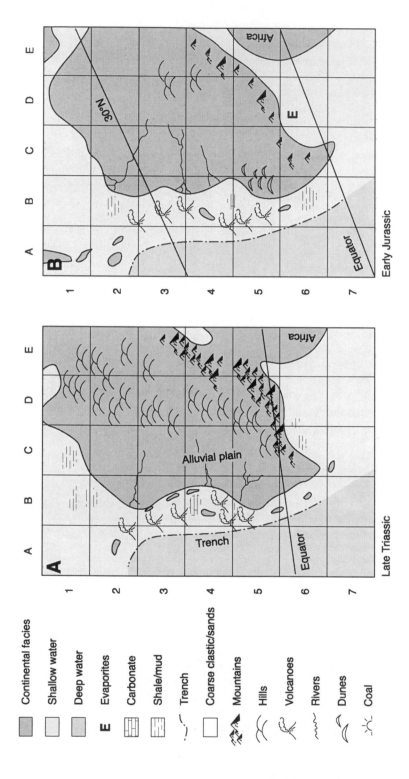

Figure 14.5   A set of four palaeogeographical maps showing the changing pattern of sedimentation and land in North America during the Mesozoic. See Figure 13.5 for locational grid and key. [Diagrams based on information from: Dott & Batten (1988) The Evolution of the Earth, McGraw Hill, Figures 16.10, 16.13, 16.19 & 16.24, pp. 472, 475, 480 & 485]

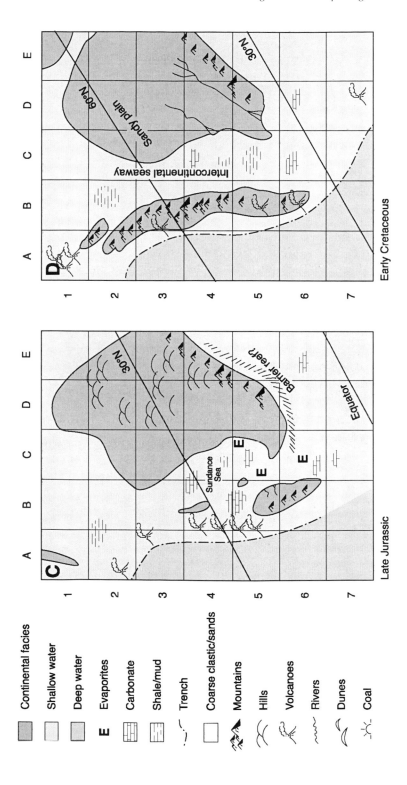

Figure 14.5 Continued

development of greenhouse conditions may have led to a global stagnation of ocean waters leading in turn to the development of widespread anoxia (see Section 10.4.1). Like the Ordovician event (Box 13.4), the Permian extinction may also have had a long lag time, with several stepped extinctions. The most obvious effects of the Permian extinction are felt outside our study area, with the greatest impact being recorded in the rocks and fauna found in present-day southern China. Here a transitional unit is present which records the ultimate demise of the reef faunas composed of ancient corals belonging to the rugose and tabulate groups. This unit is missing from most other areas, including the North Atlantic region, which were dominated by terrestrial deposits and therefore do not contain a clear record of the Permian extinction of marine groups.

In the absence of a record of the end-Permian extinction event it is often difficult to separate the basal part of the Triassic System from the upper part of the Permian within the North Atlantic region, which was dominated by desert conditions at this time (Figure 14.5). However, around the margins of our region, in western North America, and southern Europe, the presence of marine conditions and extensive carbonate deposition result in a more abundant fossil record which does record the end-Permian extinctions and the transition between the Palaeozoic to Mesozoic faunas (Figure 10.17). In southern Europe especially, carbonate-dominated shelves bordering the Tethys Ocean, which were developing along the eastern margin of central Pangaea, were a distinctive feature of this area throughout the Mesozoic.

In the Triassic, as in the Permian, desert conditions prevailed over much of the present-day North Atlantic region. In eastern North America and in northern Europe, the deposition of the terrestrial facies was controlled by fault-bounded basins, like the Newark Rifts (Figure 14.1). As discussed above, this phase of extensional tectonics, associated with the beginnings of the Atlantic Ocean, intensified throughout the Triassic. As a result a series of elongate basins, or graben, were

---

Box 14.1
The Capitan Reef Complex

The Capitan Reef complex is a Late Permian Reef complex which formed in the Delaware Basin of south-western North America (Diagram A). The reef was extensively studied by Newell et al (1953) and serves as a model for the way in which reef complexes are initiated and develop, and it also provides an interesting model for the development of a small basin which ultimately became anoxic. The reef was constructed during the Guadaloupian Stage of the Late Permian, and although its carbonates have been extensively altered to dolomite, thereby destroying much of its original fabric, it is possible to determine that the reef was actually constructed by algae, sponges and bryozoans, rather than the corals we expect today. The reef talus slope preserves many beautiful fossils which have tumbled down the slope into the basin. The Capitan complex actually consists of several reef units which have successively built outwards from the basin margin as it subsided (Diagram B). The centre of this enclosed basin became successively more anoxic, with a stratified water column of low oxygen bottom waters and oxygenated surface waters. The onset of anoxia is associated with the subsidence of the basin, as lower levels were clearly more oxygenated originally, and as the basin subsided, so the ability of freshly oxygenated waters to enter the basin was increasingly restricted (Section X-Y). Ultimately, the Delaware Basin filled with evaporites as it became more restricted, and the seas finally retreated from the Basin by the end of the Permian, leaving the reef complexes high and dry.

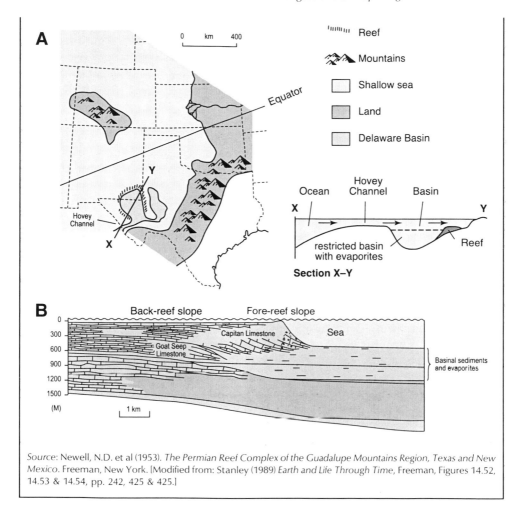

Source: Newell, N.D. et al (1953). *The Permian Reef Complex of the Guadalupe Mountains Region, Texas and New Mexico*. Freeman, New York. [Modified from: Stanley (1989) *Earth and Life Through Time*, Freeman, Figures 14.52, 14.53 & 14.54, pp. 242, 425 & 425.]

produced which parallel the margins of the North American continent today. Basin subsidence continued throughout the Late Triassic and into the Jurassic, and as a result, a thickness of almost six kilometres of red-beds were deposited, including: coarse clastic sediments eroded from the fault scarps at the graben margins and finer lacustrine sediments and evaporites deposited in ephemeral lakes. Sedimentation was interrupted by periodic intrusion of sills and dykes associated with the rifting, and as already seen, the Palisades sill along the Hudson River is an impressive example of this phenomenon. Evidence of extensional tectonics was also felt in Britain, and are manifested in the development of north–south trending graben, such as the Worcestershire Graben. As in eastern North America, Triassic sedimentation is largely confined to these graben and half-graben, the margins of which are partly defined by the Hercynian and Caledonian uplands. The Early Triassic succession in Britain is almost completely continental and devoid of marine influence.

Throughout the present-day North Atlantic region climate was hot and arid, although intense but infrequent rainfall ensured that coarse clastic sediments accumulated in alluvial fans and braided rivers, and a broad alluvial plain formed over much of the North American craton (Figure 14.5). Much of the continental area was the site of nonmarine sediments developed in an arid environment. Spectacular evidence of terrestrial sedimentation is preserved in Arizona, where the so-called Petrified Forest preserves coniferous woodland and evidence of fossil log-jams in fluvial environments. Evaporite deposits were abundant and formed in large lakes (playas) which frequently dried out to give salt pans. Reptilian remains are common throughout the Triassic rocks of our region, although they are mostly recorded by trace fossils.

Marine Triassic sediments are known from the western margin of the North American Craton, although as with the eastern seaboard, the marine influence was restricted to specific basins on the margin of the continent. In Europe, the first indications of a marine transgression from the south caused by rising sea level in the Tethyan Ocean was felt in the Triassic. In Europe these deposits consist of fine-grained clastic and carbonate sediments containing a limited marine fauna. The presence of stromatolites within these deposits suggests that this advancing sea was tidal, since stromatolites prefer such conditions. This evidence of marine transgression is confirmed by the extent of shallow marine sediments in Europe and North America in the Early Jurassic, and is characterised by the growth of epeiric seas over much of northern Europe, and a central transcontinental sea within west central North America. The marine transgression at the close of the Triassic continued on and off throughout the rest of the Mesozoic and culminated in the Late Cretaceous highstand when a vast area of the world's continents was flooded.

Towards the end of the Triassic, an extinction event occurred which is difficult to attribute to any single cause (Figure 4.10). About 20 per cent of all animal families became extinct, and for the first time a noticeable reduction in land animals is apparent (Figure 10.17). This led to the decline of the mammal-like reptiles which had dominated the terrestrial ecosystem, and opened an ecospace which allowed the radiation of the dinosaurs. There is evidence to suggest that once again the pattern of extinction was stepped, with several events, and it is not clear whether there is a direct link between the terrestrial and marine extinctions.

14.3.2 Mesozoic 'Waterworld'

The Mesozoic is a time of greenhouse conditions, which extended for longer than any other period in Earth history (Figure 10.17). As a consequence sea levels were particularly high and rose steadily during much of the Mesozoic, reaching an all-time high at the close of the Cretaceous. This is recorded by the growth of epeiric seas around the margins of Pangaea producing a kind of 'waterworld', with vast areas of continent flooded as sea levels rose. Relevant to our study area are seas transgressing Pangaea from Tethys in the east, flooding European lowlands, and from the ancient Pacific in the west transgressing the continental interior of North America. In this section we examine the outcome and impact of the development of the 'waterworld' in our North Atlantic region.

In northern Europe, the eustatic sea level rise which started in the Late Triassic culminated in the Jurassic with a flooding episode that effectively drowned much of the north-west European area of Pangaea (Figure 14.2). In North America, the development of an epeiric sea was to take longer. The continental interior remained a land area dominated by alluvial sedimentation until the middle part of the Jurassic, when the continental interior was transgressed for the first time since the Early Palaeozoic, forming the Sundance Sea (Figure 14.5). The geography of the North American continental area at this time was affected by the strong influence of its eastern margin, uplifted during the Alleghenian Orogeny, part of the Hercynian orogenic phase. In the west, subduction of Panthalassa had led to arc collisions since the Carboniferous, and marine conditions there were tempered by periodic collage collisions (Figure 14.4). Ancient geography also had a distinct effect on the epeiric sea which was developed east of the Appalachian mountain belt and ancient Greenland shield area. Here relict Caledonian upland areas, such as the Fennoscandian high, Scotland, the Lake District of England and much of Wales, and the newly formed largely Hercynian uplands of the Anglo–Brabant Landmass, the Bohemian, Rhenish and Amorican massifs in central Europe and Iberian Meseta in the west formed a series of islands, sometimes referred to as the European Archipelago (Box 14.2). This sea had its outlet to the Boreal basin north of Greenland, and to the developing Tethyan Ocean in the south-east.

Throughout much of the Jurassic sea level continued to rise, and the effects were felt widely. In Europe, Early Jurassic mudrocks were deposited uniformly over much of the flooded area by pelagic sedimentation, and these mudrocks have since proven to be an important source of oil for much of the North Sea region. These mudrocks are particularly noted not only for their potential as oil and gas source rocks, but also as conservation lagerstätten, preserving a rich fauna of soft bodied organisms in the laterally equivalent Jet Rock (England), *Posidonienschiefer* (Germany) and *Schistes de Lustre* (northern France). This fauna demonstrates the diversity of life in this productive European epeiric sea, despite its proclivity to become anoxic. This anoxia is probably related to stratification of the relatively shallow water column, or the development of a barred basin which hindered water circulation, and was to become a theme of the Jurassic epeiric sea. Locally, where the sea was shallow, limestones and ironstones were also deposited breaking the monotony of mudrock deposition. Correlation of the marine sediments of the Jurassic is excellent because of the abundance and suitability of ammonites as guide fossils within them. The marine sediments also document the dominance of molluscs, which replaced the brachiopods as the dominant marine invertebrates, following the mass extinction event at the end of the Permian (Figure 10.17).

In Europe, the progressive rise in sea level during the Jurassic was interrupted by a minor regressive phase in the mid-Jurassic (Figure 14.2). This was coincident with domal upwarping in the North Sea region and with local volcanism. Both phenomena were probably the products of crustal stretching caused by the initiation of a tensional regime, which was to lead ultimately to the opening of the Atlantic Ocean. This upwarp exposed the Lower Jurassic sediments of the North Sea to erosion and was

Box 14.2
The European Archipelago and the boundary between the Boreal Sea and Tethys

During the Jurassic and Cretaceous the north European region lay between the uplands of the eastern margin of North America/Greenland and the ancient highlands of Scandinavia. A series of islands, the European Archipelago, also dotted the sea between these land masses, and restricted the free passage of marine organisms northwards. The division of Jurassic marine organisms into Boreal or Tethyan faunas had been known since the early part of the twentieth century, but a simple distinction on climatic grounds was difficult to fathom, as there was no recognisable southern or Austral fauna which would point to an obvious bipolar distribution of organisms. Periodic interchanges and mixing between faunas were also difficult to explain using this reasoning. In 1969, Hallam suggested that the Boreal and Tethyan faunas may be restricted according to salinity changes, and demonstrated the obvious facies changes from clastics in the north of the sea, to carbonates in the south, suggesting reduced salinities in the north. Although this concept was not widely accepted, the work by Fürsich and Sykes (1977) suggested that the physical barrier created by the presence of the islands created small basins in which there was small scale environmental instability, sufficient to control the distribution of sensitive organisms. Periodic drowning of the islands during small scale sea level fluctuations would allow mixing. This example demonstrates the complexity of environments in epeiric seas

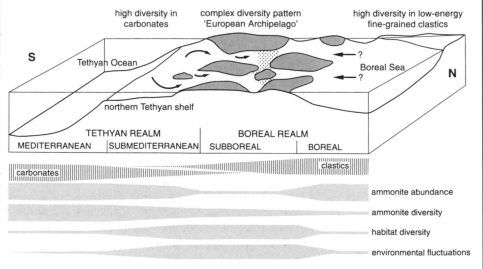

Sources: Hallam, A. (1969) Faunal realms and facies in the Jurassic. *Palaeontology* **12**, 1–18; Fürsich, F.T. and Sykes, R.M. (1977). Palaeobiogeography of the European Boreal Realm during Oxfordian (Upper Jurassic) times: a quantitative approach. *Neues Jahresbuch fur Geologie und Paläontologie, Abhandlungen* **155**, 137–161. [Modified from: Fürsich and Sykes (1977). *Neues Jahresbuch für Geologie und Paläontologie, Abhandlungen* **155**, p. 157.]

followed by controlled subsidence in this area throughout the rest of the Jurassic and Cretaceous. Coarse clastic sediments were deposited in some parts of northern Britain, particularly in north-west Scotland. Here, river deltas coalesced to form an alluvial plain, and similar environments existed in the Viking Graben of the North Sea.

Away from the North Sea region the epeiric sea decreased in depth during this phase of upwarping and several new islands emerged, including the Mid-North Sea and Ringkøbing–Fyn highs (Figure 14.2). This shallowing of the warm continental sea led to a migration of limestone deposition northwards during the mid-Jurassic. This trend was, however, not to last, as towards the end of this interval renewed transgression and sea floor subsidence gave rise to the continued deposition of mud rocks. The Late Jurassic is therefore associated with a continued increase in sea level and the extensive deposition of mud rocks, particularly the Kimmeridge Clay. This organic-rich clay is similar to that deposited in the Early Jurassic, and both sets of mudrocks provide the major source rocks for North Sea oil. This pattern is, however, complicated by continued and increased uplift and rifting, referred to as the Cimmerian Movements within the North Sea region, associated with the collision of the Cimmerian microcontinent with Asia. By the end of the Jurassic sea level began to fall, exposing a large percentage of Europe centred on the Caledonian and Hercynian uplands which had the effect of isolating southern England and a fraction of northern France from the depositional environments of northern Britain and the North Sea where the deposition of mud rocks continued. In the south, the Wessex Basin was an area of carbonate deposition at this time, which gradually transformed into a hypersaline lagoon in which coniferous forests developed in a Mediterranean-type climate.

In North America the pattern of sedimentation was also set by the development of the sea level rise documented for Europe. Marine sedimentation had been continuous for some time on the western margin of the continent, punctuated by episodes of orogenesis associated with the subduction of the Pacific plate. The development of the Sundance Sea in the Middle Jurassic (Figure 14.5) records a transgressive episode and the start of deposition of the Zuni Sequence of sediments (Box 13.3) which was to continue into the Cretaceous. The Zuni records progressive onlap onto the eroded surface of the cratonic interior, a land surface composed of dissected rocks from the Precambrian through to the Early Jurassic, which records the scale of erosion which had taken place in the continental interior, and had been free from marine influence for so long (Figure 14.5). Commencing with coarse clastic sediment, the Sundance Formation contains a range of facies types from mudrocks through to carbonates and in some cases, evaporites, which reflect a range of depositional environments experienced within this small shallow basin of the developing epeiric sea. The transgressive phase was reversed by the Late Jurassic, however, due to the continuing effect of collage collision tectonics on the Pacific margin of the continent. These contributed to the construction of the proto-Western Cordillera, and overprinted the otherwise global sea level rise, in much the same way as local events associated with the opening of the Atlantic changed the pattern of facies distributions in Europe. One effect of this was a reduction in the area of the Sundance Sea, and the deposition of non-marine sediments over the Sundance Formation around its margins. These sediments comprise coarse clastic deposits eroded from the newly emerging mountains to the west, deposited in a complex of rivers and swamps. These sediments are world famous for their vertebrate fauna, and many of the classic dinosaur finds of the early twentieth century were located within the fluvial sandstones of Morrison Formation, most famously preserved in the Dinosaur National Monument of Utah (Box 14.3).

Box 14.3
Fluvial deposits of the Morrison Formation

The Morrison Formation of central Utah is world famous for its extensive fauna of dinosaurs which has furnished many of the world's museums with spectacular examples from the Late Jurassic. The Morrison Formation consists of a succession of non-marine sediments laid down after the retreat of the Sundance Sea, and the uplift of the Western Margin of the continent during the Nevadan accretionary event. The famous pastel-coloured sands contain abundant reptilian remains, many of which are gathered together in concentrated lagerstätten, through flood events which washed the bodies of these extinct giants together. This unique assemblage gives us a clear insight into the nature of the centre of the North American continent, a vast area bounded by the Appalachian uplands in the east and the advancing uplift of the Western Cordillera in the west. Initial work by Craig et al (1955) set up the stratigraphical framework for the Morrison formation, and demonstrated the nature of the alluvial fans which preserved the dinosaur remains. Recent work by Petersen (1984) has suggested that the Cordilleran Orogeny far to the west directly influenced the environment of deposition, with small seismic shocks promoting fan deposition. The Morrison Formation remains an important hunting ground for dinosaurs and these stratigraphical studies provide the basis for understanding the background environments.

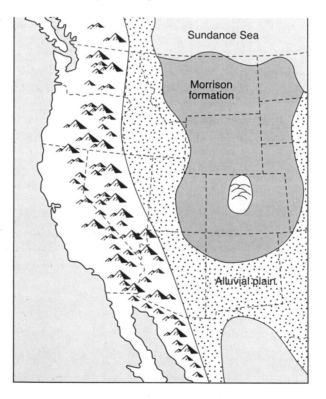

Sources: Craig, L.C. et al (1955) Stratigraphy of Morrison and related formations, Colorado Plateau region, a preliminary report. *United States Geological Survey Bulletin* 1009E, 125–168; Petersen, F. (1984) Fluvial sedimentation on a quivering craton: influence of slight crustal movements on fluvial processes, Morrison Formation, West Colorado Plateau. *Sedimentary Geology* **38**, 21–49. [Modified from: Dott and Batten (1988) *Evolution of the Earth*, Figure 16.20, p. 481.]

The control of facies patterns by relative sea level fluctuations was to continue into the Cretaceous throughout the North Atlantic region. North America and Europe were separated by the mountain belts of the Caledonian and Hercynian orogenic phases and are therefore difficult to link in a simplistic way, but the basic pattern of fluctuating epeiric seas continued throughout the Jurassic and Cretaceous. However, local controls associated with upwarps due to the onset of Atlantic rifting, as well as the development of Cordilleran collage collisions is a controlling feature of the nature of deposition in each region respectively. In Europe, marine regression at the start of the Cretaceous continued and the crustal rifting associated with the opening of the Atlantic Ocean led to uplift. At this time much of Europe was land and marine deposition was largely confined to the Tethyan margins, and to north-east England and the North Sea. In southern England and northern France uplift of the Anglo–Brabant Landmass led to the input of coarse clastic sediments into the adjacent sedimentary basins. In one of these, the Wealden Basin, a landscape of braided rivers and mudflats developed. A significant rise in sea level took place towards the end of the Early Cretaceous and rifting of the Atlantic was temporarily abated such that the eastern margin of the young Atlantic became essentially passive (Figure 14.2). This transgression brought marine conditions back to southern and midland England and northern Europe either side of the extensive land area comprising the Anglo–Brabant, Rhenish and Bohemian massifs. Mudrocks, glauconitic sandstones and some limestones indicative of a shallow sea environment were deposited over much of this area, as well as in the North Sea. The North Sea high areas were flooded for the first time during this transgression.

In North America, the pattern of sedimentation, continuing from the Jurassic, was one controlled by the continuing pattern of orogenesis on the western margin of the craton. After the regressive events at the close of the Jurassic, deposition of the Zuni Sequence continued with a mid-Cretaceous transgressive episode which drowned the location of Morrison Formation deposition and led ultimately to the Late Cretaceous Interior Seaway which at its greatest extent stretched from the Gulf of Mexico to the Arctic Ocean (Figure 14.5). The initial transgressive episode led to the widespread development of black shale facies over much of the basin centre, similar in many ways to the Jurassic black shales that had developed in the European epeiric sea. Continuing eustatic sea level rise ultimately led to the development of the extensive Interior Seaway opening into the Gulf of Mexico. For much of its early history the seaway had extensive carbonates deposited in fringing reefs around its margins, and with black shales in the centre of the basin. By the Late Cretaceous large scale depositional cycles, known as Greenhorn Cycles can be identified (Figure 14.6), with progressive deepening being indicated through the cycle of coastal sands–shales–basinal chalks, with the chalks representing the limit of the supply of clastic sediments on the margins of the seaway. By the end of the Cretaceous, new collisions on the western margin of the continent meant that uplift once more reversed the transgressive trend, leaving the cratonic interior as land once more, and closing the cycle of Zuni Sequence sedimentation.

In northern Europe the Late Cretaceous heralded the initiation of sea floor spreading in the Atlantic region (Figure 14.3), with the opening of the Rockall trough, but in general the pattern of sedimentation is that of a tectonically quiet period with

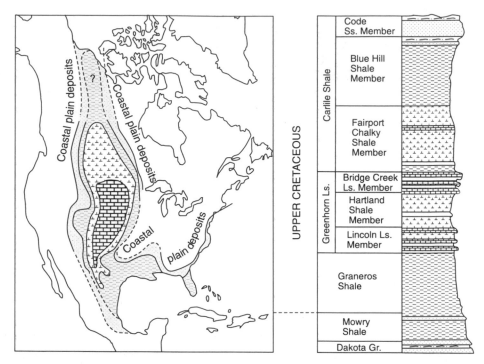

Figure 14.6 *A typical Greenhorn cycle in the Late Cretaceous Western Interior seaway of North America [Based on information in: Stanley (1989)* Earth and Life Through Time, *Freeman]*

sedimentary basins produced by simple downwarps. However, to the south, the Tethyan Ocean, which separated Africa from Europe and Asia, was closing as the components of Pangaea and particularly Gondwana, began to separate (Figure 10.2 I–J). The rise in eustatic sea level, which gave rise to the development of the Interior Seaway of North America also flooded a large part of north-west Europe, including the Anglo–Brabant Massif, leaving only a greatly reduced Central European land area, the Amorican Massif and the ancient Caledonian uplands of Scandinavia and Scotland. Over much of this area a planktonic ooze, rich in coccoliths (Figure 10.14), the calcareous remains of a marine plant, was deposited to give Chalk. The accumulation of chalk continued until the end of the Cretaceous. Tectonically, the compressive regime produced to the south of Britain by the closure of the Tethyan Ocean led ultimately to the progressive inversion of these sedimentary basins and their uplift. This, together with a major regressive phase led to the emergence and erosion of much of the younger chalk outcrop across Europe, just as the seas had retreated from the Interior Seaway of North America.

The end of the Cretaceous 'waterworld' was signalled by the onset of the regression which terminated one of the greatest sea level highs seen on Earth, demonstrating the end of the Mesozoic greenhouse. The regressive event of the Late Cretaceous is associated with a deterioration in climate which ultimately led, in the long view, to the onset of the Cenozoic 'Ice Age' (Figure 10.6). The return to non-marine sedimentation over much of the North Atlantic region coincides with what must be the most studied extinctions in the Phanerozoic history of the Earth, which saw the extinction of the dinosaurs at the end of the Cretaceous (Figure 4.10). As we have seen in Section 10.4.1 the debate still rages over the causes of this extinction, with the most likely candidates being the eruption of flood basalts, or the impact of a large meteorite, both or either of which must have contributed directly to a period of catastrophic global climate change. What is interesting is that while the extinction is pervasive amongst a wide diversity of animal groups, including large reptiles such as dinosaurs and marine organisms such as the ammonites and many foraminifera, little overall effect is recorded in landplants. This is particularly important as land plants underwent a great radiation in the Early Cretaceous as the versatile flowering plants or angio-sperms adapted to a wide range of environments. Whatever the explanation, it is clear that the discovery of the so-called 'smoking gun'—a large impact crater of the right size and age in the Yucatan Peninsula in Mexico has given a great fillip to those that favour an extra-terrestrial cause for this mass extinction (Box 10.4).

14.4   The Atlantic Opens, the Tethys Dies

Palaeogene environments of the North Atlantic region were dominated by sea floor spreading in the Atlantic, crustal thinning in the North Sea and by the final closure of the Tethyan Ocean which led to the Alpine Orogeny in southern Europe and Asia (Figures 12.1, 10.2 J–K). Sea floor spreading in the Atlantic commenced in the Late Cretaceous, but large scale activity did not commence until the Palaeogene. This resulted in widespread igneous activity either side of the diverging plate margin (Figure 14.3). Particularly important is the suite of lavas, batholiths and dyke swarms which were extruded and emplaced in north-west Britain, and which led to the regional uplift of the ancient Caledonian mountains of Scotland. The results of this volcanic activity are often referred to as the Tertiary Volcanic Province which actually forms part of a much larger episode of volcanic activity associated with the rapid opening of the North Atlantic. At the same time the North Sea was the site of rapid subsidence, such that marine conditions continued to dominate in this area during the Palaeogene. Subsidence of the North Sea was caused by cessation of thermal upwelling which had caused crustal extension in this area on and off during the Mesozoic. In fact, many would argue that the North Sea is perhaps best viewed as a rift complex which did not quite make it into a zone of plate divergence. If it had, Britain may, by now, have been located on the American side of the Atlantic. The uplift of northern Scotland during the Palaeogene provided a source of clastic debris for the North Sea basin. In contrast to, and probably as a result of, the thermal uplift in north-west Britain, southern Britain was experiencing subsidence. In conjunction with this subsidence, minor sea level fluctuations occurred, which were probably the result of variation in the rate of sea floor spreading within the Atlantic. The delicate interplay of subsidence and sea level variation in southern Britain resulted in a cyclic transgressive—regressive facies (Box 5.2). The transgressive phases within this cycle are recorded by marine sands and muds, while the regressive episodes are indicated by fluvial sands. The maximum marine advance in southern Britain took place in the Eocene, and deposited the marine muds of the London Basin, the London Clay.

On the other side of the Atlantic the Tejas Sequence is the final cratonic sequence of the North American Craton, and reflects the deposition of a broad range of facies associated with continuation of the Cordilleran Orogeny in the west, and the effects of the rifting in the east (Box 13.3). Adjacent to the Atlantic margin, the east coast of North America was receiving input of thick wedges of clastic sediments derived from the Appalachian Mountains, which had been rejuvenated during the rifting phase.

During the Late Palaeogene the effects of the Alpine Orogeny, caused by the closure of the Tethyan Ocean were felt in Europe. The Alpine Front is far to the south of Britain and comprises the Alpine, Carpathian, and Himalayan mountain belts (Figure 12.1). Large scale nappe tectonics developed throughout the Alpine Front with maximum compression in the Late Palaeogene and Early Neogene. The effects in northern Europe were restricted to basin inversion and folding of Palaeogene sediments and the gentle deformation of the Cretaceous basins in the Neogene. This was coincident with vigorous sea floor spreading in the Atlantic which caused widespread uplift in the British Isles, and this trend was reinforced by the compressional regime of the Alpine Orogeny.

14.5   Late Cenozoic Environments in Two Continents

During the Late Cenozoic sea levels were relatively low, associated with the onset of icehouse conditions (Figures 10.6 and 10.17). Consequently the Neogene sedimentary record is rather fragmentary along the Atlantic margins. In eastern England, for example, a small marine transgression resulted in the deposition of marine shelly sands. The marine molluscs of these 'Crag' deposits show that the climate was cooling rapidly towards the Cenozoic 'Ice Age' (Figure 10.6). The fossil molluscs present give valuable information on the climate as they are similar to present day molluscs which range from warm temperate to sub-arctic species. A similar pattern is developed on the east coast of North America, where Neogene deposits are found onshore in a few low-lying areas. The cliffs of Chesapeake Bay in Maryland in particular preserve evidence of a Miocene transgressive episode in a marginal downwarp. The sediments of the Chesapeake Group also record a general cooling, with a rich, temperate fauna and flora.

The cooling trend experienced in the Neogene culminated at the start of the Quaternary with the development of mid-latitude ice sheets in northern Europe and North America. These ice sheets have waxed (glacials) and waned (interglacials) in response to orbital forcing in solar radiation for the last 2 Ma and are likely to do so in the future. They are responsible for shaping the landscape we see today in much of the North Atlantic region, and have left a rich legacy of glacial landforms and sediments.

14.6   Pangaea and the Birth of the Atlantic

Coincident with the formation of Pangaea at the close of the Palaeozoic was a great change in global conditions which ultimately led to the mass extinction at the end of the Permian, itself coincident with the cessation of sea floor spreading activity and a consequent fall in sea levels (Figures 10.13 and 10.17). The North Atlantic region as we know it was at the heart of this large continent, and experienced terrestrial sedimentation with continental deserts that were to continue long into the Triassic and beyond in eastern North America. Marine deposition was restricted to the margins of the continent, and particularly bordering the Tethys, where carbonate shelves existed for the whole of the Mesozoic. Marine deposition was therefore a function of sea level variation, fluctuating within the Mesozoic greenhouse primarily due to tectonic activity at its margins, with the accretion of Cordilleran terranes in the west, and the beginning of the accretion of terranes to form south-east Asia. Flooding of the cratonic areas led to the development of extensive epeiric seas in Europe and North America, and these seas characterise the mode of deposition in the Mesozoic. Atlantic rifting ultimately led to the demise of Pangaea and greatly affected the later Mesozoic and Cenozoic world. This, together with a reverse into icehouse conditions greatly affected the mode of deposition and by the end of the Cenozoic, the familiar geographies of Europe and North America had been set.

14.7   Summary of Key Events

- **Plate Tectonics.** With the development of the large supercontinent at the start of the Permian, the tectonic regime switched from one of compression to one of extension which ultimately caused the break-up and dispersal of the supercontinent. The North Atlantic region was central to the major rift axis, and experienced periods of extensional tectonics and volcanism associated with the break-up of Pangaea. Regional uplifts created by local rifting, and by the effects of the collage collisions on the western margins of the North American continent were to influence the pattern and style of sedimentary input into what were essentially passive sedimentary systems. The closure of Tethys ultimately led to the basin inversion of many of the depositional basins within the north European context.
- **Climate.** During the early part of this interval the Earth experienced some of the warmest conditions within its history, before a gradual cooling took place in the Cenozoic which finally resulted in the establishment of an icehouse state. The period of global warmth is recorded in Britain by the deposition of extensive desert deposits, including red-beds. Within the North Atlantic region, the return to icehouse conditions resulted in the growth and decay of a succession of ice sheets.
- **Sea Level.** With the construction of Pangaea, the eustatic sea level reached an all time low. With the onset of extensional tectonics, however, the eustatic sea level began to rise steadily throughout the latter part of the Mesozoic and Early Cenozoic, reaching its high stand in the Late Cretaceous. This trend was reversed as polar ice sheets began to develop, as the Earth returned to an icehouse state.
- **Biosphere.** During this period there were several major crises within the biosphere, all of which had a dramatic impact on the faunal record in the North American and North European continental seas. The high sea levels and global warmth during part of this interval resulted in the widespread deposition of chalk at the end of the Cretaceous.

14.7   Suggested Reading

Pangaea has had a dramatic influence on the development of the North Atlantic region, and the collected papers edited by Emery et al (1994) provide a detailed introduction to this. Details of the Mesozoic to Recent stratigraphy of Britain and Europe are contained in the texts by Anderton et al (1979), Glennie (1990), Duff and Smith (1992) and Woodcock and Strachan (2000). Excellent palaeogeographic maps for the north European region are provided by Whittaker (1985), Ziegler (1990) and Cope et al (1992). The background to the development of the North American Craton, and the Western Cordillera through the Mesozoic are given in the texts by Dott and Batten (1988) and Stanley (1989). A fuller picture of North America, and particularly the development of the Western Cordillera, is given in the collections of papers edited by Burchfield et al (1992) and Gabrielse and Yorath (1992), while discussions of the Mesozoic basins are given by Sloss (1988) and Bally (1989). Important chapters dealing with the evolution of the Western Cordillera, the Atlantic Ocean, and the Quaternary history of the North American craton are to be found in the edited volume by Bally and Palmer (1989).

References

Anderton, R., Bridges, P.H., Leeder, M.R. and Sellwood, B.W. 1979. *A Dynamic Stratigraphy of the British Isles.* Allen & Unwin, London.

Bally, A.W. 1989. Phanerozoic basins of North America. In: Bally, A.W. and Palmer, A.R. (Eds) *The Geology of North America: an Overview.* Geological Society of America, Boulder, 397–446.

Bally, A.W. and Palmer, A.R. (Eds) 1989. *The Geology of North America: an Overview.* Geology of North America, Geological Society of America.

Burchfield, B.C., Lipman, P.W. and Zoback, M.L. 1992. *The Cordilleran Orogen – Conterminous United States.* Geology of North America, Geological Society of America.

Cope, J.C.W., Ingham, J.K. and Rawson, P.F. 1992. *Atlas of Palaeogeography and Lithofacies.* Geological Society Memoir No. 13.

Dott, R.H. and Batten, R.L. 1988. *Evolution of the Earth.* (Fourth Edition) McGraw-Hill, New York.

Duff, P. McL. D. and Smith, A.J. 1992. *Geology of England and Wales.* Geological Society, London.

Embry, A.F., Beauchamp, B. and Glass, D.J. (Eds). 1994. *Pangea: Global Environments and Resources.* Canadian Society of Petroleum Geologists Memoir 17.

Gabrielse, H. and Yorath, C.J. 1992. *Geology of the Cordilleran Orogen in Canada.* Geology of Canada, Geological Survey of Canada.

Glennie, K.W. 1990. *Introduction to the Petroleum Geology of the North Sea.* (Third Edition) Blackwell, Oxford.

Sloss, L.L. (Ed.) 1988. *Sedimentary cover – North American Craton: US.* Geology of North America, Volume D2, Geological Society of America.

Stanley, S.M. 1989. *Earth and Life Through Time* (Second Edition). Freeman, New York.

Whittaker, A. (Ed.) 1985. *Atlas of onshore sedimentary basins in England and Wales: post-Carboniferous Tectonics and Stratigraphy.* Blackie, Glasgow.

Woodcock, N.H. and Strachan, R.A. (Eds) 2000. *Geological History of Britain and Ireland.* Blackwell, London.

Ziegler, P.A. 1990. *Geological Atlas of Western and Central Europe.* (Second Edition). Geological Society, London.

15

Summary of Part Two:
The Pattern of Earth History

In Part Two, we have examined the concept that the different depositional environ-ments on Earth today are primarily a function of the interaction of four key variables: plate tectonics, climate, sea level and biosphere. Together, these variables control not only the nature of depositional basins, but also the type and preservation of sedimentary facies contained within them. Through the application of uniformitarian principles, it is clear that the Earth's stratigraphical record actually reflects the nature and distribution of depositional environments through time. As such, it follows that the stratigraphical record is primarily a product of temporal changes and the nature of interaction in the key variables.

In Chapter 10, we explored the global rhythm of change within each of the key variables. The changing pattern of continents and plates can be viewed as a large cycle—the supercontinental cycle—in which the Earth's continents have been first dispersed and then assembled into a large supercontinent. The Phanerozoic records one such cycle, and there may have been at least two more in the Precambrian. The Earth's climate has oscillated throughout the Phanerozoic between periods of global refrigeration (icehouse states) and periods of global warmth (greenhouse states). Sea level has also fluctuated during this time with rises and falls responding to both tectonic and climatic patterns of change. Finally, the biosphere has developed through geological time as life on Earth has evolved and diversified, periodically suffering extinctions. Part Two therefore illustrates that the global pattern of the interaction of the key variables provides a framework upon which the detail of the geological development of an area can be built.

In the final chapters of this book, we have examined the geological development of one small portion of the Earth's crust with respect to this record. Our chosen area was the North Atlantic region, although we could equally well have selected almost any continental area that preserves a suite of sedimentary rocks. The North Atlantic region has been well-studied by generations of geologists, and in the continents of Europe and North America many of the basic principles of geology were born. Yet, despite this coverage, new ideas and concepts continue to be developed to explain the development of this region, and from the mid-1960s the concept of an earlier Atlantic, now known as the Iapetus, has helped shape our ideas about the development of orogenic belts. A long record of passive-margin sedimentation was initiated at the close of the Hercynian orogenic phase, and continued up to and including the opening of the Atlantic Ocean in the Mesozoic. Examination of this adequately demonstrates that once plate tectonics has done its job in developing basins for sedimentary accumulation, then the lesser controls of sea level, climate and biosphere will shape its contents. What is clear, is that in no one area of the Earth's crust are all the global events recorded. For example, although generations of geologists may have identified the presence mass extinctions in the Permian and Triassic preserved in marine sediments, these are absent from the continental sediments deposited at this time in Europe and North America. Despite this, it is true to say that the stratigraphy of the North Atlantic region is an extraordinarily rich one, and will repay the attention of any observer, with many areas still awaiting fuller investigation.

As an example of this concept, the Precambrian record of the North American Craton is revealing the role of progressive accretion of Archaean and older crustal elements, with significant growth in the Proterozoic. Although opinions differ on the nature of Precambrian plate tectonics and orogenesis, a fuller record is emerging of a great continental heartland at least 3800 Ma old. At a different scale, the Late Proterozoic to early Phanerozoic record of the North Atlantic region is fascinating, as again plate tectonic processes are of paramount importance. Those of us in Britain used to viewing stratigraphy as if, metaphorically, from the wrong end of a telescope are used to the difficulties of presenting a coherent whole to our story, particularly as the most ancient rocks in Scotland appear to be developed in fault-bounded terranes. Only by reversing our metaphorical telescope can we hope to determine the influence and impact of the key variables of Earth history. By taking a broader perspective it is possible to see our region as the centre of a vice-like closure between two great cratonic regions (Laurentia and Baltica), and that a defining orogenic phase, the Caledonian was to shape and develop northern Europe and North America, leaving the record we know today. Later events were important, but perhaps of a lesser extent than the development of Pangaea, a defining interval in stratigraphy.

Using Pangaea, it is possible to examine the nature of the interplay of the four main variables of Earth history, and to see and identify the difficulties of the approach. Simply put, the return to a supercontinent after the Proterozoic one—Palaeopangaea or Rodinia—not only had a direct role in both creating and destroying sedimentary basins, but also carried that role forward in developing the interplay of climate—sea level—biosphere. If we take the Late Palaeozoic–Early Mesozoic completion of Pangaea as a turning point, then it is possible to examine the early Palaeozoic, and to a lesser extent, Late Mesozoic records as products of orogenesis, swept up in the

Caledonian accretionary deformation phase, and the Mesozoic Atlantic opening. A cumulative history of successive Wilson Cycles created Pangaea. At a time when the Earth was still in icehouse mode, the cessation of sea floor spreading activity with the construction of the supercontinent meant that the sea level fell. Climate had little role to play here, and was overprinted by the bigger picture of changing ocean basin volume. Ultimately, of course, this was probably to play a major role in the mass extinction at the close of the Permian, as habitat space in shelf environments was stretched to the limit, and the continental interior was baked, recorded by the widespread deposition of terrestrial arid facies.

Traditional accounts of stratigraphy have attempted to build up a pattern of Earth history brick-by-brick, layer-by-layer. This obscures the true importance of global change in the development of the stratigraphical record. We have attempted to provide an alternative approach, stressing the importance of the global interaction of the key variables through time, but equally recognising the nature of the local environment in determining local stratigraphies. In Part Two we have provided a framework of global change upon which, we hope, you will be able with the tools presented in Part One, to reconstruct the nature of changing environments through time for any area of the Earth's crust, represented by the stratigraphical record.

Lexicon of Stratigraphical Terms

The following is a lexicon of the most important stratigraphical terms used in this book. In it each term is defined and the first major usage of the term in the book is identified by a page number. Terms given in italic are for cross-referencing purposes.

Absolute dating: dating of rocks in years, primarily based upon the decay of radioactive elements in minerals and organic matter (*radiometric dating*), 55.

Accommodation space: volume of a basin available for deposition, 71.

Accretion: see *crustal accretion*.

Accretionary prism: triangular wedge of sediment built up in slices along the outer edge of a trench during *subduction*, 113.

Acme biozone: *biozone* based on the total abundance of a *guide fossil* species, 41.

Actualism (see *uniformitarianism*): concept of the uniformity of process in developing the Earth's stratigraphical record. As a statement of scientific method it assumes no process or action can be admitted as having acted to produce a geological unit except those which we can observe operating today, 9.

Adaptive radiation: interval of rapid expansion and diversification of organisms on Earth, often following a *mass extinction*, 180.

Allocthonous terranes: see *terranes*.

Allocyclic controls on sedimentation: external influences controlling sedimentation in a given basin, 69.

Angular unconformity: relationship between two stratified rock units, in which the lower is tilted and eroded, and the upper unit, deposited on the erosion surface and forming the unconformity, is lying at a lower angle, 21.

Assemblage biozone: *biozone* based on the combined ranges of several *guide fossil* species, 41. Sometimes known as an Oppel Zone.

Asthenosphere: plastic layer of the Earth's mantle underlying the lithosphere, 104.

Autocyclic controls on sedimentation: internal influences controlling sedimentation in a given basin, 69.

Back-arc basin: *sedimentary basin* to the rear of a *volcanic island arc* on *convergent plate margins*, 114.

Barrow Zones: zones of metamorphic *facies* defined by the presence of characteristic minerals, demonstrating particular pressure and temperature conditions, 64.

Batholith: large deep-seated and irregular body of intrusive igneous rocks, 239.

Biofacies: facies recognised on palaeontological grounds, 96.

Biogeography: see *palaeobiogeography*

Biosphere: the sum total of all life on Earth at any given time, 173.

Biostratigraphy: branch of stratigraphy that involves the study of fossils to enable correlation of sedimentary successions, 36. The principal unit of biostratigraphy is the *biozone*, which is determined using *guide fossils*.

Biota: group of organisms, 211.

Biozone: strata organised into a stratigraphical unit on the basis of their fossil content, 41. Recognition of biozones requires the selection of appropriate *guide fossils*. Principal types of biozone are: *acme biozone*, *assemblage biozone*, *consecutive range biozone*, *partial range biozone* and *total range biozone*.

Cambrian explosion: rapid diversification of organisms with hard parts at the opening of the Cambrian, associated with continental flooding and nutrient enrichment due to changes in oceanic circulation, 185. Also referred to as the Cambrian expansion.

Catastrophism: eighteenth century concept that the Earth's geological record can be interpreted as a sequence of catastrophic events, 8.

Chemostratigraphy: branch of *event stratigraphy* involving *correlation* of rock sequences on the basis of their chemical composition, 49.

Chronostratigraphical Scale: global standard of chronostratigraphical rock units (e.g. *systems*), 51.

Chronostratigraphy: global standard stratigraphy, consisting of global rock units (e.g. *systems*) based on *stratotype* sections arranged into a standard scale by international agreement, 51.

Collage collision: mountain building through collision of many plate elements or fragments of ocean floor (*ophiolite*), 118. Collages are made up of *terranes*, each with their own stratigraphical history.

Consecutive range biozone: *biozone* defined on the consecutive ranges of evolving *guide fossil* species, 41.

Conservative plate margin: plate margin in which two *lithospheric plates* move laterally past each other along a transform fault, 107.

Continental collision: production of mountain chain from the closure of an ocean basin and the collision of two continents along its margin, 116.

Continental drift: movement of the continents through time, a function of *plate tectonics*, 103.

Continental margin orogenesis: production of mountain chain where oceanic crust is subducted beneath continental crust, leading to compression, 116.

Convergent plate margin: plate margin in which *lithospheric plates* are in relative motion towards each other, 106.

Correlation: comparison of stratigraphical units in space and time with others in different areas, 26.

Craton: area of continental crust, 147. Ancient cratons are sometimes referred to as shields, after the shape of the North American craton.

Cross-cutting relationships: method of determining *relative chronology* through examination of such cross-cutting features as faults, folds, intrusions and metamorphism, 20.

Crustal accretion: continental growth through the accretion of smaller crustal entities during plate tectonics, 146.

Cyclothem: cyclical arrangement of facies, commonly associated with sea level variation, particularly in the Carboniferous, 156. Known as Yoredale facies in the UK.

Décollement surface: decoupling of the upper layers of the Earth's crust from the lower layers during deformation, 238.

Diachronism: concept that units of uniform lithology may have formed at different times and in different places, 45.

Diastem: obscure or subtle time gaps of uncertain origin that occur in a stratified rock sequence, 23.

Disconformity: relationship between two stratified rock units, in which the lower is eroded, and the upper unit, deposited on the erosion surface and forming the unconformity, is lying the same angle of dip as the lower one, 21.

Divergent margin: plate margin in which *lithospheric plates* move apart with the formation of oceanic crust, 104.

Dropstones: anomalous large clasts in a fine grained setting, 157. These have been commonly associated with ice rafting, but may be the result of other processes, such as rafting of clasts in tree roots for example.

Dyke: steeply inclined or vertical sheet like igneous intrusion, 122. Valuable in defining *cross-cutting relationships*.

Earth history: history of the Earth revealed from its geological record. Relies upon the interpretation of rock units as sequences of events through time, 1.

Ecospace: habitats available for colonisation, 218.

Eustatic sea level: global sea level variation, controlled by changes in the volume of ocean basins, or changes in the volume of the oceans themselves, 71.

Event horizons: geological products of events, such as that produced by a volcanic eruption, 46. Event horizons are time significant.

Event stratigraphy: correlation of rock sequences using the products of events (*event horizons*) in the stratigraphical record, 46. *Magnetostratigraphy* and *chemostratigraphy* are branches of event stratigraphy.

Exposure: that part of an *outcrop* visibly exposed at the Earth's surface, 26.

Facies: Sum total of the characteristics of a rock body that indicate its environment of deposition, 63. For sedimentary facies, typical characteristics are: geometry, lithology, structure, and fossil content.

Facies analysis: analysis of facies in order to define a palaeoenvironmental model for a given set of rocks, 63.

Facies fossil: fossils of limited value as *guide fossils* due to their dependence on particular *facies*, 37.

Faunal and floral succession: evolutionary concept that successive species of animals and plants replace each other in time, recorded as fossils in the stratigraphical record, and valuable in *biostratigraphy*, 35.

Fore-arc basin: *sedimentary basin* in front of a *volcanic island arc* and behind the trench on *convergent plate margins*, 114.

Foreland Basin: basin created through loading by advancing thrust sheets (*nappes*), 114.

Formation: a unit of largely homogeneous lithology that can be clearly denoted on a map, 26. Formations may be linked into groups, and subdivided into members on the basis of lithology.

Geopetal: fossil 'spirit-level', a *way-up criterion* formed by the partial filling of voids in fossils, 15.

Geosyncline: outdated concept of the development of fold mountain belts from the subsidence of linear troughs or basins of sediments adjacent to the margins of continental cratons, 102.

Glacial: intervals of ice sheet expansion during an *icehouse* period, 158.

Golden spike: figurative spike 'driven in' at the internationally agreed boundary separating chronostratigraphical units, 228.

Gradualism: eighteenth century concept that the Earth's geological record can be interpreted as a sequence of gradual events, proceeding at a uniform rate, 8.

Graphical correlation: semi-quantitative method of *correlation* of rock sequences, 43.

Greenhorn cycles: large scale depositional cycles in the Cretaceous of North America, with alternation coastal sands-shales-basinal chalks, 269.

Greenhouse: period of global warmth with much reduced or absent polar ice sheets, 154.

Greenstone belt: belt of Precambrian metasediments and metavolcanics probably associated with ancient island arcs, 199.

Guide fossil: fossils useful in the *correlation* of sedimentary rock units, and which define *biozones*, 37. Also known as zone or index fossils, they are selected according to a set of ideal criteria, outlined in Figure 4.1.

Hallam sea level curve: plot of *eustatic sea level* through time based on continental flooding, 73. Used as an alternative to the *Vail curve*.

Hyposmetric curve: a plot of land surface areas at different elevations valuable in determining *relative sea level* variation, 71.

Icehouse: period of global cooling characterised by the formation of high latitude ice sheets, 154.

Ichnology: the study of *trace fossils*, 95.

Index fossil: see *guide fossil*.

Interglacial: intervals of ice sheet minimum or warmth during an *icehouse* period, 158.

Interpretative tools: tools used in the interpretation of rock units as the products of events in Earth history, 2.

Island arc: see *volcanic island arc*.

Lagerstätten: finds of exceptionally preserved (conservation Lagerstätten) or exceptionally concentrated (concentration Lagerstätten) fossil assemblages, 95.

Lithodemic unit: lithostratigraphical term for non-stratified rock units, such as igneous intrusions, 26.

Lithosphere: rigid upper layer of the Earth's mantle, 104.

Lithospheric plate: thin, rigid plates in motion over the surface of the Earth, 104.

Lithostratigraphy: the formal description of rock units and their comparison in space and time, 25. The standard currency of lithostratigraphy is the *formation*.

Log: record of stratigraphical observations at a given locality, and used in *correlation*, 29.

Macroevolution: the broad pattern of evolution, including the origin and diversification of major groups, 175.

Magnetostratigraphy: branch of *event stratigraphy* involving *correlation* of rock units on the basis of their magnetic properties, usually polarity reversals of the Earth's magnetic field, 48.

Mantle plumes: bodies of molten rock originating in the Earth's mantle, 109.

Mappable unit: see *formation*.

Mass extinction: rapid changes in the standing diversity of life on Earth over and above the background levels of extinction, 176. Several causes have been postulated, including changes in sea level and climate, environmental changes associated with vulcanicity, and the impact of large extra-terrestrial bodies.

Microcontinent: small crustal entity, 146.

Microevolution: changes within organisms up to and including the creation of new species, 174.

Milankovitch cycles: cyclic changes in the amount of solar radiation received by the Earth caused by rhythmic changes in the Earth's orbit, 160. These changes affect climate and have left a record in the Earth's sedimentary record. Three orbital cycles are recognised: (1) eccentricity of the orbit (95 000 to 400 000 year cycles); (2) axial tilt (obliquity; 40 000 year cycles); and (3) precession of the equinoxes (21 000 year cycles).

Nappe: sheet of deformed rocks bounded by thrust faults, 114.

Nonconformity: relationship between two rock units, in which a lower, crystalline unit is eroded, and an upper, stratified unit is deposited on the erosion surface, forming an unconformity, 22.

Oceanic trench: trench created by the *subduction* of one *lithospheric plate* beneath another, 113.

Offlap: progressive movement of marine *facies* offshore during *regression*, 70.

Onlap: progressive movement of marine *facies* onshore during *transgression*, 69.

Ophiolite: fragment of ocean crust that has been preserved through obduction, a process associated with subduction in which ocean floor is thrust up and welded on to a continental margin, 211.

Oppel Zone: see *assemblage biozone*.

Orogenesis: process of mountain building, 116. Four processes are involved: *continental margin orogenesis, continental collision, collage collision* and *transpression*.

Orogenic belt: linear belt of mountains created by *orogenesis*, 199. These belts were formerly called *geosynclines*.

Orogeny: interval of mountain building, 116.

Outcrop: total lateral extent of a geological unit as it occurs at the Earth's surface, 26. Outcrops differ from *exposures* in that they may be obscured by surface deposits or human development.

Outcrop map: map of the total lateral extent of a geological unit as it occurs at the Earth's surface, 26.

Overlap: progressive *onlap* of marine *facies* onshore during *transgression*, with the most landward facies lying directly upon the land surface, 69.

Overstep: progressive *onlap* of marine *facies* over a folded and eroded basement during *transgression*, thereby creating an angular unconformity, 69.

Palaeobiogeography: study of the geographical distribution of organisms of the ancient Earth, 136. Often shortened to biogeography.

Palaeoclimatology: the study of the climate of the ancient Earth, 85. Data can be gathered from stable isotopes, lithology, fossils and computer models.

Palaeoecology: the study of the interaction of organisms with each other and with their environment in the geological past, 92.

Palaeogeography: study of the geography of the ancient Earth, 81.

Palaeomagnetism: relict magnetic properties of some rocks, providing a means of determining *palaeogeography*, and valuable in *event stratigraphy*, 82.

Paraconformity: significant time gap, the result of prolonged non-deposition, between two stratified rock units lying the same angle of dip, 23.

Parasequences: *onlapping* or *offlapping* units that make up *sequences*, 75.

Partial range biozone: *biozone* based on the partial range of a *guide fossil* species, 41.

Passive continental margins: trailing edges of continental margins resulting continental rifting and *sea floor spreading*, 112.

Phyletic gradualism: pattern of microevolution involving gradual change, 175.

Plate tectonics: concept that the surface of the Earth consists of a series of thin, rigid *lithospheric plates* in motion, 104.

Plume tectonics: the periodic upwelling of large bodies of molten rock (plumes) within the mantle, and their interaction with the *lithosphere*, 109. Sometimes referred to as pulsation tectonics.

Province: term used in the context of Precambrian cratons, defining areas bounded by orogenic belts, 199.

Pulsation tectonics: see *plume tectonics*.

Punctuated equilibrium: pattern of microevolution involving of no change (stasis) punctuated by rapid intervals of change, 175.

'Punctuated gradualism': concept that evolution proceeds through a combination of gradual evolution (*phyletic gradualism*) punctuated with rapid shifts in morphology, 175.

Radiometric dating: dating of rocks in years, based upon the decay of radioactive elements in minerals and organic matter, 57.

Realm: term used in biogeography and *palaeobiography* to denote area delimited by the distribution of organisms, on a broad scale, 136.

Regression: progressive retreat of the sea from the land during sea-level fall, 69.

Relative chronology: relative order of formation of rocks in a given area, 17.

Relative dating: relative order of formation of rocks in a given area, 17. Principal tools are *biostratigraphy* and *event stratigraphy*.

Relative sea level: concept that sea level has varied through geological time, 69. Sea level variations may be local or global (*eustatic*), and may lead to *transgressions* or *regressions* of the sea over the land surface.

Rift valley: *sedimentary basin* formed during the earliest stages of the Wilson Cycle, 111.

Right way-up criteria: see *way-up criteria*.

Rock-preservation curve: graph defining the amount of rock preserved as a percentage of the maximum through time, 125.

Rock-stratigraphical units: units of *lithostratigraphy*, such as *formations*, which are not time significant, 51.

Sea floor spreading: process in which plates are spread apart at a mid-oceanic ridge, and which drives the opening of ocean basins, 103.

Sedimentary basins: sites of accumulation of sedimentary rocks, 110.

Sedimentary facies: see *facies*.

Seismic exploration: geophysical exploration of sedimentary sequences, often offshore, using seismic waves, 74.

Seismic reflector: distinct boundaries between distinct sedimentary units identified during *seismic exploration*, 75.

Sequence: use informally for the sedimentary succession at any given locality, an formally as a package of sedimentary rocks bounded by *unconformities*, 75. Each sequence is composed of separate *parasequences*, and *sequence boundaries* have a special significance, often controlled by *allocyclic* mechanisms.

Sequence boundary: boundary between successive *sequences*, unconformable onshore and often conformable offshore; sequence boundaries are of great significance in *sequence stratigraphy*, 79.

Sequence stratigraphy: study of *unconformity* bounded sedimentary *sequences*, and application of this study in correlation, 79. Sequence stratigraphy has been influential in the development of a new method of considering the stratigraphical record.

Sequencing tools: tools of *relative chronology* in stratigraphy, 2.

Shield: see *craton*.

Sill: horizontal or subhorizontal sheet of intrusive igneous rocks, 251.

Strata: rock layers, generally of sedimentary or volcanic origin. Singular: stratum, 1.

Stratigraphy: formally, the study of stratified rocks. More generally, it is the key to understanding the Earth's crust and its materials, structure and past life, 1.

Stratotype: locality selected by international agreement representing the best exposed representative of the boundary between *chronostratigraphical units*, 53. Boundary stratotypes are denoted by figurative *golden spikes* driven in at the boundary point.

Subduction: plate tectonic process in which one plate dives down beneath another and is consumed at depth, 106. Leads to the production of an *oceanic trench*.

Successor basin: sedimentary basin formed along the line of suture between two convergent continental masses, 234.

Supercontinent: large amalgamation of smaller continental blocks through *crustal accretion*, 146. Several supercontinents are known from the Earth's geological record, including Rodinia (Palaeopangaea) in the Proterozoic and Pangaea in the Palaeozoic, and their formation and dispersal is seen to form a *supercontinental cycle*.

Supercontinental cycle: periodic formation (by *crustal accretion*) and dispersal of *supercontinents*, 146.

Superplumes: large bodies of molten rock that originate at the junction of the core and mantle, 109.

Superposition: principle that states that in any undisturbed stratified sequence, the layer at the bottom of that sequence was formed first, 11. In most cases, *way-up criteria* are required to prove superposition.

Suspect terrane: see *terrane*.

System: standard unit of *chronostratigraphy*, defined as bodies of strata formed during specific intervals of geological time. These units have time-significant boundaries, and are arranged into a global *chronostratigraphical scale*, by international agreement, 52.

Terranes: mappable structural entity with a stratigraphical sequence distinct from its neighbour, the result of *collage collisions*, 118. Such entities are often described as allochthonous terranes in that they may have moved some considerable distance before accretion. Suspect terranes is a term used in the initial identification of such entities.

Thermohaline circulation: pattern of deep ocean circulation driven by differences in the salinity of the Pacific and Atlantic oceans, 163.

Tie lines: lines of *correlation* linking lithological equivalency in two or more rock sequences, 28.

Time lines: lines of *correlation* linking time equivalency in two or more rock sequences, 28.

Time-stratigraphical units: units of *chronostratigraphy*, such as *systems*, which have time significant boundaries, 51.

Time tools: tools of *relative chronology*, and specifically *relative* and *absolute dating* tools, in stratigraphy, 2.

Total range biozone: *biozone* based on the total range of a *guide fossil* species, 41.

Trace fossils: the tracks, trails and feeding traces of fossil organisms, 95.

Transgression: progressive advance of the sea onto the land during sea-level rise, 69.

Transpression: deformation along a conservative or transform margin, 119.

Transtensional basin: sedimentary basin formed by extension on transform faults, 114.

Unconformities: relationship of rock units defining a time gap between the formation of one rock unit and the formation of another. This gap can be short or long. Principal unconformity types include: *angular unconformities, disconformities, nonconformities, paraconformities* and *diastems*, 21.

Uniformitarianism (see *actualism*): encompassing the dictum that 'the present is the key to the past'. A general principle guiding the interpretation of rock units as events in the geological history of the Earth, by the study of processes occurring today to interpret the products of the past, 7.

Vail curve: *eustatic sea level* curve derived from the seismic study of sedimentary *sequences*, 81.

Volcanic island arc: arc of volcanic islands caused during the *subduction* of a *lithospheric plate*, 106.

Walther's Principle: concept which demonstrates that as environments shift through time, so the respective sedimentary *facies* of adjacent environments will succeed each other in a vertical profile, 67. As such, in a stratigraphical sequence with no breaks, the vertical profile is effectively equivalent to the lateral facies variation at any one time.

Way-up criteria: structures (sedimentary, volcanic, palaeontological and tectonic) that provide a record of the original way-up of a rock sequence, thereby proving original *superposition*, 15.

Well-logs: log of the stratigraphical units in a borehole, and used in *correlation*, 29.

Wilson Cycle: life cycle of an ocean from opening to closure, 109.

Wire-line logging: *log* of the geophysical observations made by a measuring device lowered down a borehole or well, and used in *correlation*, 29. Typical geophysical techniques are listed in Table 3.2.

Zone fossil: see *guide fossil*.

Index